Urban Sustainability

The Urban Sustainability Book Series is a valuable resource for sustainability and urban-related education and research. It offers an inter-disciplinary platform covering all four areas of practice, policy, education, research, and their nexus. The publications in this series are related to critical areas of sustainability, urban studies, planning, and urban geography.

This book series aims to put together cutting-edge research findings linked to the overarching field of urban sustainability. The scope and nature of the topic are broad and interdisciplinary and bring together various associated disciplines from sustainable development, environmental sciences, urbanism, etc. With many advanced research findings in the field, there is a need to put together various discussions and contributions on specific sustainability fields, covering a good range of topics on sustainable development, sustainable urbanism, and urban sustainability. Despite the broad range of issues, we note the importance of practical and policy-oriented directions, extending the literature and directions and pathways towards achieving urban sustainability.

The series will appeal to urbanists, geographers, planners, engineers, architects, governmental authorities, policymakers, researchers of all levels, and to all of those interested in a wide-ranging overview of urban sustainability and its associated fields. The series includes monographs and edited volumes, covering a range of topics under the urban sustainability topic, which can also be used for teaching materials.

Ali Cheshmehzangi

Australia's Urbanity

Architecture, Place, and the Pulse of Urban Life

Springer

Ali Cheshmehzangi
School of Architecture, Design
and Planning
The University of Queensland
Brisbane, QLD, Australia

ISSN 2731-6483 ISSN 2731-6491 (electronic)
Urban Sustainability
ISBN 978-981-95-3880-5 ISBN 978-981-95-3881-2 (eBook)
https://doi.org/10.1007/978-981-95-3881-2

This Springer imprint is published by the registered company Springer Nature Singapore Pte Ltd.
The registered company address is: 152 Beach Road, #21-01/04 Gateway East, Singapore 189721, Singapore

Home is not a place, but a state of being — where the soul feels it belongs!
Photo taken in May 2025, by Ali Cheshmehzangi.

To the First Nations peoples of Australia.
With deep respect for your enduring connection to land, sky, and waters;

For your knowledge systems that have shaped this continent for tens of thousands of years;

For your resilience in the face of injustice, and your wisdom in the face of silence.

May this work honour your custodianship, your stories, and your right to a just and thriving future.

We walk on Country with gratitude, and we acknowledge that sovereignty was never ceded.

Always was, always will be Aboriginal land.

Preface

"Life is neither a destination nor a race; it is the moment you breathe and create purpose, meaning, and connection—with yourself, with others, and with the world around you".

We hardly realize how our cities influence us as we move through them on a daily basis. We talk about houses, roads, parks, and buildings as though they are apart from who we are. However, we tend to overlook that cities are more than just our homes. They are manifestations of our grief, paradoxes, dreams, customs, memory, and culture. The city is a reflection. It reveals the things we value and disregard. It serves as a reminder, frequently too late, of how we have prioritized spectacle over substance, quantity over quality, speed over spirit, development over purpose, and destruction over advancement. Buildings and traffic patterns are only one aspect of what we refer to as "urban life". It is the theatre of our entire existence.

Nevertheless, we have been completely wrong about this for the majority of the past century. People were not considered when we planned our cities. Then, we ended up with so many development-based cities rather than purpose-driven planned cities. Without knowing the earth beneath them, we have constructed skylines. We have exported lifestyle models that conceal identity. Our urban frameworks have become so globalized that a city on one continent might replicate another halfway around the world, disregarding factors like location, climate, people. culture, memory, and significance. Our cities have become monotonous and empty as a result of our all too frequent exaltation of sameness as modernity. We have destroyed local wisdoms, sabotaged deep localities, eliminated vernacular knowledge, and disregarded the pulse of what made our communities genuinely alive in our unrelenting quest for efficiency and international or global appeal.

Architecture, once a language of belonging, has turned mute. The profession has lost its way in its obsessions with formality, scale, and superficial "innovation" due to its long-standing attraction to monumentality. Architecture has been hailed over the past century as artifice rather than manifestation of quality, spirit, and values, and as corporate aspiration rather than social duty. Both intentionally and unintentionally, we have eroded our social license to define what architecture must be. In chasing aesthetics, prestige, and profit, we have drifted from architecture's deeper purpose, i.e., to serve people, to honour place, and to uphold collective dignity. What was once a civic art rooted in human experience and cultural continuity has, too often, become a vehicle for exclusion, spectacle, and market-driven narratives. As a result, we now face a crisis not just of design, but of trust; a rupture between the built environment and the communities it was meant to uplift. Like many other countries, Australia has absorbed these practices without questioning whether they are beneficial to us. Students have been trained to imitate "iconic" models and to look for status in form rather than meaning. However, that story is not carried further in this book. It is vehemently questioned in this work.

Here, we ask: What is the true pulse of urban life? Why is architecture more than just a tangible medium? What is meant by "designing with and for place"? What does it mean to pay attention to the layers of language, land, connection, and memory that come before and will outlive us? These are not rhetorical questions. They are urgent.

In our fractured and worn-out contemporary world, which is so dominated by speed, anxiety, and disconnection, we are realizing that we have fundamentally misunderstood what it means to live well. The symptoms of this are everywhere in cities: loneliness, isolation, housing crises, environmental degradation, and the erasure of community. Rather than being the result of poor planning, this is a collapse of values; we have forgotten what it means to belong, not just to a place, but also to each other and to time. Until we regain that sense of belonging, no amount of "smartness", no glossy towers, and no effective transportation grids will be able to restore what we have lost.

It is here that Australia has something profound to offer. This continent, the land of the oldest continuing civilization on Earth, holds a wisdom that modernity has neglected, but can no longer afford to ignore. The knowledge systems of Aboriginal and Torres Strait Islander people are alive, breathing philosophies of place, existence, and stewardship, not just cultural artifacts to be conserved. I just hope we can revitalize and bring those valuable knowledge back to the core of our thinking and design, as well as in our design thinking, architecture, placemaking, and co-creating life. Rhythms, not deadlines, are what they teach us. Not about supremacy, but about coexistence. Listening to earth, water, wind, and fire as family rather than as resources to be exploited. None of this information is "alternative" or "complementary"; rather, it is necessary.

This exploitation has only changed over time; it has never ceased but only evolved. Industrialization sped it up, excelled it to a new level, and the delusion of deindustrialization covered it up. To be clear, however, deindustrialization never really happened. The purpose of the deliberate rebranding was to hide the continuous resource looting, the never-ending cycle of needless production and consumption, and the ravenous need for the so-called "progress". Industrialization did not end; rather, it was merely softened by language and approved by the same old – if not middle-aged – white-collar men who have dominated world affairs in recent centuries. These are the same deceptive and cunning architects that reframe bribery as express services, land grabbing as colonization, extermination as self-defence, genocide as combating terrorism, occupation as annexation, cultural erasure as democracy, and test beds as experimentation. We continuously witness tragedies though, those that are failures of our so-called civilizations we brag about. Yet, every word and linguistic change is a deliberate act of dishonesty and disgrace. So, when are we going to wake up? When will we realize that globalization is our manufactured downfall, not our shared future? A slow-moving doomsday, cloaked in the reassuring rhetoric of development and marketed to us as progress. It's time we demanded a change and named it what it is. This cannot last forever; otherwise, we won't last much longer.

But amid this long deception, there are wisdoms we have ignored, wisdoms that never relied on industrial conquest or linguistic manipulation, but on deep-rooted connection, humility, and care, wisdoms held by the world's Indigenous communities, who have long known a different way to live, lead, and sustain. The profundity of Indigenous thinking, not only in ecology and the environment but also in leadership, governance, health, and culture, has humbled me on my travels through Australian towns and villages. For example, listening comes before acting, consensus comes before ego, and service comes before title in Aboriginal leadership concepts. These are not just noble ideals. They are incredibly useful strategies for creating communities and cities that promote healing rather than damage. They serve as a reminder that unless a place recognizes all its inhabitants and layers of history, it cannot be considered fully created.

But the most heartbreaking part of my travels was the haunting "absence" of Aboriginal populations, the original custodians of these majestic lands and waters under a clear and pure sky of many stars; the many beautiful stars that we now rarely see in a full spectrum, perhaps to avoid letting us see and know the world beyond our existence in this egoïstic living. The presence of the Aboriginal communities, once vibrant and sovereign, has been silenced by many years of erasure, barbaric annihilations of many rounds, and sabotaging attempts of "closure". They were not merely displaced; they were forcibly exiled from their homes, communities, and homelands, torn from Country with unspeakable cruelty, and made strangers on their own soil. All these, I can hear in every pulse of the lands and communities I passed through. What should have been a memorable journey through shared history instead became a quiet confrontation with emptied landscapes, heavy with the grief of stolen lives and civilizations deliberately demolished, disregarded, and discarded. I could—and can still—sense, hear, and connect with the enduring forced silence emerged from the beforehand bellowing, lamenting, and those deep and everlasting crying outs with symbolizing torture, sorrow, and agony.

This is beyond just a minimal dismantlement; this is an obvious crime against a knowledgeable and caring nation. Their absence echoes louder than any monument: a deep tragic indication to a legacy of harsh and uncivilized invasion, intentional dispossession, and systematic forgetting. Yet the silence that I experienced in parts of my travels stays very vigorously in my brain, their stories endure so deeply in my heart, etched in landforms, whispered by winds, and remembered by those who still fight to respect, recognize, and revive their truth. To regain all that, the oppressed need to be respected against all these inequalities and oppression, the resistance must be cherished, and the false beliefs need to fade away.

It is not the people of this land that need to be forgotten, nor the wrong doings against them shall be ever forgiven. It is the people of this land that need to be at the pulse of every heartbeat of these lands and waters, and the failures of the colonizers in respecting the place, the people, and the spirit of Country must be named, remembered, and reckoned with. Until justice breathes where silence once settled, and until their rightful presence is honoured (with genuine intentions, and not with artificial acts, fabricated wordings, and deceptive gestures) in every policy, every place, and every story told, the land itself remains "incomplete".

Indigenous societies have been practicing what we now refer to as "sustainability" for thousands of years, without much fanfare or hashtags; not only here in Australia, but everywhere else, too. In places where the locals and their locality are replaced by new and globalized values, it becomes painfully clear that we have failed profoundly and irreversibly. We know without a single doubt that Aboriginal civilizations have experienced intricate and flexible systems of land management, seasonal migration, and communal care—what we now refer to as "regenerative design". In songlines, stories, and ceremonies, they exemplify what we now refer to as "resilience". They are not outdated methods of knowing. They are highly relevant, connected, and complex. What Aboriginal people have always considered "smart" is rooted in deep listening, ecological reciprocity, intergenerational responsibility, and living with the land, not above it. Intelligence, for them, was never measured by data flows, concrete towers, glass facades, or algorithmic speed. It was measured by how well one cared for Country, remembered stories, respected kinship systems, and ensured that the land thrived long after one's own lifetime. In stark contrast, what the contemporary Western world proudly brands as *smart development* is often soulless and predisposed technology or technologies wrapped in blatant and unsustainable progress—glossy infrastructures disconnected from place, automated systems that ignore human wisdom, and surveillance cities that monitor but do not understand. If smartness leads to social detachment, ecological destruction, and cultural erasure, then it is not smart at all. It is merely mechanized arrogance. True "smartness" comes from knowing when not to build, when to give back, and when to listen. That is the intelligence we've ignored, and that is the failure we must urgently confront. Thus, we must know our "modern" or "contemporary" lifestyles, which are fast-paced, fragmented, consumerist, and extractive, are actually the ones that are out of rhythm. The past does not have to catch up with us. We must make up for it.

This book is a call to realign. It is a manifesto for reconsidering how we view cities and ourselves, not a list of initiatives. Through 25 case studies from all throughout Australia, we demonstrate urban possibility as well as urban development. These are tales of innovative place-based design, architecture that respects the soil it rests on, and design that listens before acting. Every story adds to a broader discussion about how to firmly establish important values like humility, connection, caring, and dignity as the foundation of urbanity.

Let us state unequivocally that Australia has contributed to urban knowledge beyond the last two centuries of colonial urbanism. For more than 65,000 years, this continent has been pulsing. Action, not just words, is required to honour that tradition of collective thriving, layered narrative, and caring for the country. It is time to construct more intelligently, listen more intently, and design more respectfully.

Imitation of the past is not encouraged here. The idea that the knowledge of the oldest cultures in the world will influence how we live in the future is provocative.

Australia's Urbanity is not just about cities. It is about *us*. About our lives, our relationships, and our search for purpose; how we live, how we connect, and how we find meaning. Planners, designers, academics, and residents who still think that cities can be more than marketplaces, that streets can hold (and tell) stories, and that structures can preserve memories should read this book. It is a book that opposes silence. That's a bold statement. That means we need to do more for ourselves and our built environments.

We cannot afford to get this century wrong. Another century of alienation, placeless design, and rootless urbanism would come at too high a cost. Let's turn at this point. Let this book serve as a reminder of what really counts. Let this be a giving that is influenced by the knowledge of the past, the pressing need of the present, and the optimism for the future.

Welcome to *Australia's Urbanity*. May it move you. May it provoke you. And above all, may it reconnect you to architecture, to place, and to the living, breathing pulse of urban life.

Brisbane, Australia Ali Cheshmehzangi

Acknowledgements This book is the result of many minds, voices, and communities coming together, each contributing to the evolving story of urban life in Australia. I wish to express my deepest gratitude to those who have made this journey possible.

First and foremost, I extend heartfelt thanks to *Professor Sue Harrison*, at that time Dean of the Faculty of Engineering, Architecture and Information Technology (EAIT) at The University of Queensland, for her persistent support, encouragement, and belief in the importance of critical and grounded urban discourse and deep knowledge creation. Her support has provided both the space and inspiration needed for this work to grow. I also sincerely thank *Professor Deborah Terry*, the Vice-Chancellor and President of The University of Queensland, for her visionary leadership and unwavering commitment to excellence and impact across disciplines. I particularly would like to thank her for her trust in and support of my leadership, which has been instrumental in enabling many initiatives at the School of Architecture, Design and Planning (ADP) at UQ. We have strived to succeed not only in academic excellence, but also in shaping a culture of innovation, collaboration, and purpose-led transformation. "Create Change" is not just a motto we follow at UQ; we want to ensure we are fully committed to the city, the region, the state, and country. We are pursuing bold ideas, elevating our community impact, and positioning ADP as a leading force in creating the next generation of healthy, resilient, and sustainable futures.

I am also profoundly grateful to the many individuals and communities across Australia who have shared their stories, insights, and experiences to a person who has just arrived at this wonderful country. Somehow deep inside, I feel more connected to this land than I imagined I would. Perhaps, I have been here before in my previous lives. That I sometimes think about when the wind moves through the trees just so, or when the red earth crumbles softly under my feet, as if whispering an old story I almost remember. In those quiet moments, I don't just feel like I belong here, I feel as though the land remembers me too.

It is in the lived realities of place-making, neighbourhood resilience, and everyday encounters that the true pulse of urban life is found. A big and special thank you to the community leaders, advocates, elders, residents, planners, architects, and students who continue to shape our cities with care and conviction. You are the backbone of our collective progress and the true architects of change.

This book is offered as a humble contribution to the ongoing dialogue about what "*Australian urbanity*" means, and what it can become.

Competing Interests The author has no competing interests to declare that are relevant to the content of this manuscript.

About the Book

Australia's cities are undergoing a profound transformation. The way people live, work, and interact is changing as urban environments change due to the interaction of public spaces, architecture, and economic expansion. Through an extensive collection of case studies that showcase the country's most ambitious urban undertakings, ***Australia's Urbanity: Architecture, Place, and the Pulse of Urban Life*** explores this progression.

The book depicts the dynamic changes that characterize contemporary Australian urbanity, from the reimagining of public areas and waterfronts to the growth of major business districts and the emergence of high-density dwelling. It looks at urban infill projects that strike a balance between progressive redevelopment and heritage preservation, master-planned communities created for long-term sustainability, and economic innovation precincts that propel technological growth. The book presents a convincing picture of how cities are adjusting to changing societal demands, environmental issues, and population increase through these five topics.

At the heart of this exploration is the question of resilience and sustainable development, how can Australian cities evolve without losing their identity? The case studies show that combining sustainable infrastructure, smart design, and public involvement is essential for a successful urban change. While many key initiatives demonstrate the need of sustainability and innovation in forming future communities, some development-oriented projects demonstrate how urban renovation can improve both economic vitality and social inclusion. The book ends with a discussion of Australia's urban future, highlighting the necessity of planning frameworks and policies that strike a balance between ecological responsibility and expansion. *Australia's Urbanity* is a guide to comprehending the forces that will define the cities of the future, not merely a mirror of the present.

- **Showcases iconic urban projects** shaping Australia's evolving cityscapes and public spaces.
- **Explores sustainability and innovation** in master-planned communities and economic precincts.

- **Blends architecture, history, and policy** to reveal the pulse of Australian urban life.
- **Features 25 compelling case studies** from waterfront redevelopments to high-density CBD expansions.
- **Offers a forward-looking perspective** on resilience, growth, and the future of Australian cities.

Contents

About the Author

Ali Cheshmehzangi is a Professor and Head of the School of Architecture, Design and Planning (ADP) at The University of Queensland. He is also an Adjunct Professor at the School of Architecture and Urban Planning at Hunan University. He is the Brisbane Olympics Committee member for Games Legacy and Growth Sub-Committee as well as the Expert member of the 2032 Games led by UQ. He has been in the World's top 2 percent field leader since 2021, recognised by Stanford University. He is among the top 20 global scholars in the urban sustainability research area.

Ali has a wealth of experience, expertise, and a strong commitment to innovation and excellence in the fields of architecture, design, and planning, the three core disciplines of his School at UQ. With a career spanning over two decades, he has made significant contributions to the academic and professional communities, with a focus on sustainable and environmentally conscious design. Prior to joining UQ, Ali held several strategic leadership and senior managerial roles, such as Vice-President for International Engagement and Global Partnership, Founding Director and Head of the Center for Innovation in Education and Research, Head of the Department of Architecture and Built Environment, Founding Director of the Urban Innovation Lab, Director of a university-wide Teaching and Learning platform, Founding Director of International Research Network for Rural and Urban Development, Founder and Head of Research Group for Sustainable Built Environment, co-director of university-wide research priority areas, Director of Center for Sustainable Energy

Technologies, and Interim Director of Digital Design Lab. So far, Ali has published over 500 journal papers, articles, conference papers, book chapters, and reports. This is his 44th academic book, alongside his many poetry collections that he has written in Parsi/Persian and English. Some of his previous academic books have received awards at the national, provincial, and municipal levels. He also has received international awards and recognition for his research on urban resilience studies and sustainability research, as well as a 2018 Vice-Chancellor's award for his impactful contribution to higher education.

Chapter 1
Australia's Urban Pulse: Charting the Life of Cities Through Transformations, Experimentations, and Urban Transitions

1.1 Australian Urbanity: Shaping Places, People, and Possibilities

Australia's cities pulse with life. They are alive with changing seasons, lively streets, unique neighbourhoods, and well-established communities. The world may have less knowledge about the great design-based approaches in Australian cities. Beneath their well-known urban landscapes and recognizable monuments, however, is a more complex tale: a continuous one about sustainability, place-making, identity, belonging, and the potent interaction of architecture, people, and the environment. To tell this story, this book delves deeper than just the architecture and aesthetics of Australia's built environments, examining their profound power to influence people's lives, create connections, and motivate change.

At its heart, this book advances that urbanity is more than just a physical characteristic; beyond density, infrastructure, or the quickening of metropolitan borders. As discussed below, urbanity is a shared experience and a conversation between people and place that is influenced by environmental conditions, design, culture, and history. In Australia, urbanity encompasses our tensions and aspirations, our conflicts and inventiveness, and, most importantly, our capacity to create meaningful lives within intricate urban ecosystems.

The Australian urban condition is undeniably unique. Australia's urban experience is both condensed and broad, in contrast to many other countries. The nation's urban fabric, which is concentrated along beaches yet stretches into the outskirts of suburbs, reflects both openness and intimacy, past patterns of settlement, and current demographic changes. Australian cities are full of contrasts: busy inner-city marketplaces contrast with calm green corridors, peaceful residential areas contrast with global skylines. Driven by a delicate balance between urban density and the immense natural landscapes that characterize the nation's topography, these tensions are essential components of comprehending the Australian urban identity.

A. Cheshmehzangi, *Australia's Urbanity*, Urban Sustainability,
https://doi.org/10.1007/978-981-95-3881-2_1

Central to Australia's urban identity is the concept of place. The Australian notion of place-making aims to create a true sense of belonging and deeper social ties rather than just improving an area's appearance or functionality. This is quite unique in compared to other countries. Australia's histories, stories influenced by colonial legacies, migratory waves, Indigenous custodianship, and continuous community-building initiatives, are what give its cities life. Urban areas provide as potent platforms for cultural expression, recollection, and healing, as evidenced by the nation's overt dedication to multiculturalism, reconciliation, and cultural celebration.

Architectural and design excellence, not only as technical mastery but also as essential "social agency", is a critical component of our urban inquiry. In Australia, architecture, planning, and urban design frequently go beyond simple practical or aesthetic concerns to actively impact social dynamics, health outcomes, and the well-being of communities. To actively involve communities in forming their surroundings, inclusive and participatory design approaches are becoming more and more important in Australian urbanism. Architecture may be a force for good through creative design techniques and thoughtful interventions, promoting cultural integration, enhancing social cohesiveness, and producing areas that support communities rather than just house them.

Another important axis of the Australian urban experience is sustainability and resilience. Heatwaves, floods, bushfires, and droughts are among the increased challenges that Australian cities confront as a result of climate change, and they call for an inventive and aggressive urban response. From wide green spaces and nature-based solutions to energy-efficient architecture and adaptive reuse of existing infrastructure, integrated sustainability practices are becoming more and more central to Australia's urban development. We can argue that Australia's planning philosophy is currently based on resilient urbanism, which is radically changing how governments, corporations, and communities approach upcoming city-building and revitalization projects.

Yet, the objective of this book is not just to draw attention to difficulties and complications. Instead, it looks for chances to take deliberate action, encouraging civic engagement and urban change. It involves expressing potential, acknowledging accomplishments, taking lessons from failures, and handling uncertainty with clarity, wisdom, and hope. The case studies presented throughout this book, spanning diverse contexts and typologies, illustrate not only how Australian urbanity shapes collective experiences but how it offers pathways for future engagement, creativity, and shared purpose.

In charting the pulse of Australia's urban life, this book ultimately seeks to provoke conversations about the kind of urban futures Australians aspire to create (Bose, 2024). It raises important queries: How can Australian cities continue to be creative but culturally rooted, prosperous but egalitarian, and lively yet inclusive? In what ways can urban development simultaneously promote social peace, environmental resilience, and collective identity? By offering knowledge, motivation, and useful frameworks that define urban development as moral, social, and cultural obligations rather than just as economic or infrastructure activities, this book seeks to assist readers in navigating these difficult questions.

By doing this, this book presents an ambitious vision for the urban destiny of Australia, one that is characterized by deliberate design, mindful growth, sincere community involvement, and a dedication to building metropolitan areas where everyone can prosper. It is an exploration of how Australia's urban pulse can influence, mould, and improve people's lives while making sure our cities continue to be resilient, lively, and genuinely meaningful places to live.

1.1.1 Urbanity Redefined: Beyond Cities as Spaces

The idea that urbanity encompasses more than just urban growth or the actual existence of cities is presented in this section. It presents urbanity as the core of Australians' social, cultural, and environmental interactions with their natural and constructed surroundings. It draws attention to Australia's distinctive urban dynamics, which are influenced by its large landmass, varied climate, diverse population, and multiculturalism. Here, the concept of urbanity encompasses more than just growth and infrastructure; it also refers to the nature and character of human interactions, the distinctiveness of locations, and the complex interplay between urban and rural settings in forming Australian life.

When people talk of urbanity, they frequently picture high-rise towers and skyscrapers, congested city streets, and busy metropolitan areas. However, this book offers a deeper interpretation of urbanity that goes well beyond its outward manifestations. The core of how communities interact and relate with their built and natural environments on a social, cultural, and environmental level is embodied by Australian urbanity. It transcends the conventional boundaries set by infrastructure and population alone, symbolizing the dynamic interaction between the three P's of "people, place, and purpose".

Australia's urban fabric has been greatly impacted by its distinct geography, which is broad, expansive, and frequently severe (Buxton, 2014). Australian cities and towns exhibit a complexity derived from their relationship with nature and spatial dynamics, with urban populations concentrated along coastal corridors and extending inward across expansive suburban and regional areas. Australians' unique sense of place and identity is shaped by the conflict between urban concentration and natural expansiveness. In this Australian setting, urbanity is not just determined by architectural form or population size; rather, it arises from a greater understanding and reverence for indigenous heritage, natural landscapes, and community ties.

Australia's urbanity is further defined by its diverse population, which adds layers of cultural richness. Vibrant mosaics of multiculturalism, cities like Brisbane, Sydney, Melbourne, Perth, Adelaide, and Hobart celebrate diversity in their architecture, public areas, markets, festivals, and daily life (Cheshmehzangi et al. 2025). By impacting the conception, design, and habitation of urban environments, this cultural diversity improves urban experiences (Cheshmehzangi et al. 2024). It makes it possible for cities to become vibrant, living representations of the cultural diversity of the country by fostering inclusive conversations about identity, belonging, and social cohesiveness.

Nonetheless, Australia's urbanity also recognizes the close relationship between urban and rural settings. I got to experience this by driving from Melbourne to Geelong, tracing the coast, then veering inland toward Canberra. I took it as a rare opportunity to experience the rhythm of Australian urban life not from a desk or a map, but from their relationships with people and environment, the region, rural areas, and communities that I stopped by and studied. With each stop in small towns or rural pockets and communities, I began to see the shades of Australian urbanity unfold in unexpected ways. It was in the everyday routines of locals, the design of public spaces, the quiet pride in main streets, and the informal interactions in cafés and civic centres that I came to understand what urbanity truly means in the Australian context. All that had to happen not by just flying into cities and meeting project owners, but to experience them with the communities and living through them for hours or days.

From Canberra, the journey stretched further and towards Sydney. From one capital to another, a political hub to a financial center. Then through coastal towns and growing regional centres of New South Wales and Queensland, each place offered a distinct rhythm, an evolving identity, a story of community adaptation. Finally, I reached Brisbane, finishing a 2,300-km expedition that spanned seven days and dozens of necessary and examining stops. Along the way, I engaged in conversations with residents, observed patterns of urban form and function, and gathered insights that no dataset alone could offer. These places had to be experienced, and in between them, I had to think thoroughly cruising through 40–50 m tall trees alongside the quiet streets. I avoided highways for a good purpose, to allow me to see and observe better, have regular stops in places that matter, and understand the urbanity of each community I passed through. It was a short journey in time, but a rich journey in meaning. One that taught me the diversity, resilience, and layered character of Australia's urban pulse, which was a great opportunity for me to know more before writing this book. The journey also confirmed that the best way to understand a country's urban future is to move through it, slowly, attentively, and with curiosity. Every single stop mattered, and I got to know the realities on the ground, not on books, papers, maps, and on my computer. This meant a lot for a person who is relatively new to the context, understanding the depth of communities as an urbanist, interacting with everyday life as a visitor, and observing places, environment, and purpose of each location on the way up to Brisbane.

As I travelled, it became more and more evident to me that Australian urbanity is not isolated; rather, it recognizes its continuous connection to regional and rural places. That continuity meant a lot, for a country of such long-lasting civilization, with the values that are yet to be discovered as we move along. Australia's urban state is characterized by an ongoing conversation between the city and the countryside, in contrast to many other parts of the world where urbanization frequently denotes a break from rural life. Cultural exchanges, ecological ties, and economic flows that affect communities and landscapes are all shaped by this mutual reliance.

Rural communities and regional towns are essential components of Australia's larger urban and metropolitan environments; they are not peripheral (Carson et al. 2022). This is a debatable topic but that is how we can observe them as part of a unified ecosystem. They provide alternate lifestyle options based on environmental stewardship, community resilience, and a strong sense of place identity. These locations serve as a reminder that vibrancy is not limited to urban areas by showcasing slower, more personal, and frequently more sustainable types of urbanity. Their ongoing impact casts doubt on the urban-rural dichotomy and supports the notion that Australia's cities are interwoven components of a wide, dynamic, and diversified national fabric rather than discrete hubs of growth.

In this way, Australian urbanity is layered; it is as much about what happens in the heart of Sydney or Brisbane as it is about life in Wagga Wagga, Toowoomba, or Bellingen. Each informs the other, shaping a shared future where urban development and rural vitality evolve together, not apart. Thus, urbanity turns into a gauge of how people interact with one another, how places allow for deep connections and encourage diversity, creativity, and resilience. It embraces ideas of equality, social participation, and ecological responsibility, going beyond simple functionalism or economic productivity. Australian cities celebrate life quality, environmental sustainability, and community well-being by fostering human potential in addition to providing space for expansion.

Lastly, my journey in March 2025 helped me to rethink and recapture the realities with more depth and knowledge. To me, rethinking urbanity encourages Australians to consider more carefully what actually contributes to the success of cities. It urges communities, planners, architects, and legislators to take an intentional approach to urban development, giving ecological harmony, cultural diversity, and human-scale interactions top priority. Australia has the potential to lead a global movement towards more sustainable, inclusive, and compassionate urban living by recognizing that urbanity is essentially about the nature and quality of interactions; i.e., between people, places, and surroundings. By doing this, urbanity is exposed as a powerfully transforming force that has the power to reshape Australian cities into significant and meaningful places where everyone can thrive and feel like they belong.

1.2 Place Identity: Australia's Distinctive Urban Narratives

Cities tell stories. History, geography, and local values have all influenced and precast the stories that are told in their streets, buildings, and public areas. Urban environments in Australia have a particularly strong narrative power, expressing a multi-layered complexity rooted in colonial histories, Indigenous heritage, immigration waves, and changing community aspirations. Hence, Australian cities, towns, and precincts are more than just places of business; they are representations of communal identity that take the form of unique urban narratives that promote belonging, connection, and remembrance.

There is a persistent dedication to conserving Indigenous heritage at the core of Australia's urban narratives. Every town and city in Australia is situated on land that is rich in traditional knowledge, culture, and country care. This is incredibly important so that we can feel, understand, and navigate through the realities of many wonderful and valuable generations that shaped our communities as well as the soul and spirit of the nation. Understanding this basis has become essential to contemporary place-making techniques, impacting urban planning through cultural precincts, public art, storytelling, and landscape treatments that respect and celebrate Indigenous identities and histories. These initiatives and projects ought to be beyond just the sentimental gestures and towards meaningful and purpose-led decision making in urban design and planning. This involvement actively rewrites the nature of urban areas, making sure that their stories are inclusive, honest, and significant, rather than just being symbolic. The presence of Indigenous people in urban areas provides Australians with an essential, continuous conversation about respect, reconciliation, and cultural sustainability, turning urban spaces into sites for introspection and education.

Australia's distinctive urban narratives are further shaped by geography. The natural environment has a significant impact on place identity, from coastal cities that cling to dramatic shorelines to interior communities that are characterized by rivers, mountains, valleys, and desert heat. Adapting to this varied terrain and creating spaces that creatively and sensitively respond to local climates and landscapes is what Australian urbanism is all about. Geography firmly establishes urban and regional identities in concrete and accessible ways, whether in the unique alleyways of Melbourne, the verdant subtropical parks of Brisbane, the vibrant harbourside of Sydney, or the isolated tenacity of Alice Springs. This adaptability makes each metropolitan area feel distinctively Australian and suited to its particular context, while also fostering authenticity and strengthening community attachment.

However, community values, i.e., the way people interact with and impact their built environment, have an equal role in shaping place identity in Australian cities and urbanism. Communities, planners, designers, and legislators engage in ongoing discourse to shape urban narratives. Community-driven development, collaborative urban regeneration, and participatory place-building are all examples of Australia's dedication to elevating local voices and making sure that urban settings represent shared goals. By establishing thriving precincts that embody common ideals such as sustainability, diversity, inclusivity, and resilience, this strategy turns cities and towns into dynamic manifestations of social life.

In addition, Australia's urban narratives exhibit incredible resiliency and creativity. Australian towns have a history of embracing change and adapting to global trends, economic cycles, social upheavals, and environmental calamities. Adaptive reuse, careful regeneration efforts, and creative design techniques that maintain character while allowing for new uses are common ways that urban precincts change. This is also embedded in our education at my School where we give the opportunity to each student to understand the context in depth, critical thinking, and a design-based approach to architectural and urban solutions. Outdated infrastructure is converted into green spaces or active precincts, heritage areas are revitalized as cultural districts, and former industrial sites are turned into thriving community centers. Every change opens a new chapter in the urban narrative, demonstrating Australia's ability to honour its past while fearlessly embracing the future.

With each layer of historical experience, environmental responsiveness, cultural involvement, and community-driven action, place identity in Australian urbanism is thus always deepening and changing. Australia's urban narratives serve as a reminder that towns and cities are dynamic representations of shared histories, aspirations, and collective identities rather than just being physical places. Australia's embrace of place-making as a key component of urban development places its cities as potent platforms for social cohesion, cultural interchange, and storytelling—living examples of the country's inventiveness, resiliency, and ability to reinvent itself over time.

1.3 Architecture as Social Agency: Designing for Connection

This section foregrounds architecture and urban design as potent instruments of social engagement and civic transformation. In Australia, the built environment is viewed as a purposeful and dynamic medium that fosters diversity, human connection, and collective life rather than just being a physical structure. Reimagined as an active force that influences daily encounters, cultural manifestations, and community emotional landscapes, architecture serves as a social catalyst. The story emphasizes the critical role that local stakeholders, architects, and urban planners play in collaborating to create environments that strengthen the social fabric, promote diversity, and build a sense of belonging. Australian urbanism demonstrates its ability to foster meaningful interactions and an inclusive public life through careful spatial design and contextual adaptation.

In Australia's urban context, architecture is not simply a matter of aesthetic or structural design; it is a form of social agency. Every building, structure, road, and public area shapes how people relate to, engage with, and identify with their environment. An intentional approach to architecture makes it an active participant in community life, influencing social behavior, encouraging inclusivity, and fostering a place's civic identity. The foundation of contemporary Australian urban thinking is this view of design as a means of transformation and connection.

Cities in Australia are increasingly recognising that architecture must create spaces that are not only functional or aesthetically pleasing but also egalitarian and socially dynamic. According to this perspective, design choices have a big social impact. It is expected of architects and urban planners to think about how spatial layouts might facilitate interactions between strangers, provide refuge and comfort, and act as forums for cross-cultural dialogue. Architecture may purposefully create inclusive and linked spaces, whether it is through the design of a community center, the hospitable openness of a civic plaza, or the smooth transition between private and public areas in a housing development.

The development of public areas in Australian cities is among the most obvious examples of this idea in action. Parks, libraries, transit hubs, and urban plazas are now viewed as essential municipal resources rather than merely functional spaces. These areas are intended to accommodate the unofficial customs of city life, such as play, celebration, protest, rest, and conversation. In Australia today, good public architecture seeks to foster contact between people of all ages, socioeconomic levels, and cultural backgrounds while also lowering social barriers. The city's human experience is directly influenced by architecture, whether it be through interactive features, ample seating, shade structures, or Indigenous design principles.

This commitment to social agency is also apparent in how Australian architects approach housing and neighbourhood design (McGaw and Pieris 2014). It is extremely difficult to provide inexpensive, socially integrated housing in a nation with one of the most suburbanized populations in the world. We are witnessing an increasing number of new developments that prioritize human-scale connection through pedestrian layouts, integrated community facilities, and shared courtyards. Rethinking how Australians coexist is possible with medium-density initiatives, co-housing concepts, and build-to-rent programs. Architecture acts as a mediator, promoting civic engagement and neighbourly ties while protecting privacy and dignity.

Crucially, Australia's multiculturalism influences architectural reactions that honour individuality rather than standardize it. Through material selections, symbolic allusions, or heritage-honouring programming, buildings in several precincts around cities like Melbourne, Sydney, and Brisbane represent the cultural specificities of their communities. Here, architecture takes on the role of cultural narrative, promoting inclusivity while recognizing diversity. By ensuring that spaces are developed from inside community beliefs and histories rather than being imposed, this culturally responsive design approach strengthens social agency.

Additionally, architecture has agency in institutional, health, and educational contexts. University campuses are evolving from academic enclosures to cultural and intellectual commons (Boal 1998), hospitals are now designed with biophilic principles to encourage healing and emotional wellbeing, and schools are being reinvented as community anchors. Architecture affects the ambiance, usability, and emotional tone of the spaces where people connect, learn, and heal in each of these areas. I cannot name any other place in the world that can have all of these in one place. Hence, I thought sharing these examples and success stories will help Australian architects, designers, and planners to reach further, and for other places to reach Australia and experience the urban urbanity that is unique, influential, and authentic.

All these changes are the result of a change in professional perspective. In Australia nowadays, architects and urban designers play a more multidisciplinary and cooperative role. Instead of being the lone form-maker, the architect today collaborates with planners, landscape architects, Indigenous custodians, community organizations, and government organizations as a listener, facilitator, and co-creator. This collaborative ethic guarantees that architecture is a shared process based on empathy, communication, and accountability rather than a remote imposition (Wasserman et al. 2000). Ultimately, to describe architecture in Australia as social agency is to confirm that architectural form is never neutral. Every design decision, whether it concerns materials, movement, visibility, or access, either supports or contradicts a location's social dynamics. Architecture has the ability to elevate, bring people together, and change the environment in which everyday life takes place. According to this perspective, Australian architecture is a significant civic endeavour that centers urban life around people, relationships, and dignity.

1.4 Sustainability, Resilience, and Adaptation: Australia's Urban Imperative

The pressing necessity to adapt is increasingly defining Australia's urban identity. The idea of urbanity, which is at the heart of this work, needs to be reinterpreted through the prisms of sustainability, resilience, and adaptation as climate change picks up speed and environmental dangers increase in frequency and severity. In Australian cities, these are now necessities rather than optional factors. Speaking of Australian urbanity today means discussing how cities adapt to, deal with, and change because of ecological constraints, disruptions, and crises.

The challenges are severe. One of the most climate-exposed countries in the world is Australia. Cities and regions have experienced disastrous floods, protracted droughts, record-breaking heatwaves, and devastating bushfires in the last ten years alone. From sweltering suburbs and overstretched water supplies to crumbling infrastructure and precarious housing, these climate extremes have exposed the weaknesses of our urban surroundings. As a result, the idea of urban resilience is now central to Australian city planning and architecture.

In the Australian context, resilience is more than just disaster recovery; it is about increasing the physical, ecological, and social systems' ability to withstand shocks and adjust to change. This covers how cities manage their water and energy infrastructure, plan for future expansion, provide public areas, and interact with their citizens. For example, green infrastructure is being incorporated more and more into cities like Brisbane to control flooding and create liveable urban microclimates. The urban heat island effect (UHIE) is being lessened in Melbourne through the implementation of tree canopy techniques and climate-adaptive building rules. These instances are indicative of a larger national trend in which environmental awareness and climate literacy are ingrained in urban planning and governance.

Sustainability, too, is now deeply woven into the language of Australian urbanity. It is no longer limited to technical definitions of waste reduction or energy efficiency. Instead, sustainability is seen as a comprehensive idea that includes environmental care, economic viability, and social equality. This is seen in Australian cities through increased support for compact, walkable precincts, improved infrastructure for public transportation and cycling, building adaptive reuse, and initiatives to lessen reliance on fossil fuels. In order to build spaces that promote wellbeing, lessen their impact, and help future generations, local authorities, architects, and communities are working together on sustainable design projects that go beyond compliance.

Australian urbanity is also increasingly shaped by a dual awareness of fragility and opportunity. On the one hand, cities have to deal with the consequences of previous design, which have increased the environmental danger to certain areas. These consequences include ecological deterioration, social inequality, and suburban sprawl that is centered on cars. On the other hand, cities are evolving into research hubs for adaptation and resilience. Australian cities are experimenting with new lifestyle models that directly address ecological boundaries, ranging from precinct-scale renewable energy systems and regenerative landscape planning to urban agriculture and water-sensitive design.

Crucially, these changes represent a more profound cultural and philosophical reorientation than merely technological fixes. A resilient urbanity prioritizes connection over consumption, caring over control, and slowness over pace. It welcomes ambiguity with humility and forward-thinking with accountability. This philosophy is becoming more prevalent in Australian urban discourse, where listening to Country, respecting Indigenous knowledge systems, and acknowledging the interconnectedness between human settlements and the natural environment are all becoming more and more important.

Essentially, resilience and sustainability are now inextricably linked to the larger Australian urbanity effort. They outline the types of cities that Australia will require in the ensuing decades, not merely to survive, but to flourish. Urban design, architecture, and planning are being mobilized to future-proof communities, lower systemic risk, and promote new forms of urban life based on balance and belonging, as this book's case studies illustrate. More importantly, the urban pulse of Australia is pulsing in time with the country's shifting environment. However, it can also beat in harmony with the nation, the community, and a sustainable future if it has vision, integrity, and creativity.

1.5 Urban Futures in Australia: Towards Purpose-Led Planning and Integrated Urban Design in Placemaking and Precinct-Based Development Plans

Australia stands at a critical juncture in its urban evolution. How successfully the country adopts place-based strategies, integrated urban design, and purpose-led planning will determine how its cities develop in the future as it faces the twin challenges of population increase and climate unpredictability. To integrate long-term objectives with the daily realities of communities and the natural systems that support them, this vision of urban futures must go beyond reactive development and disjointed policies. Instead, it must be intentional, inclusive, and transformative.

Planning with a purpose (or purpose-led planning) entails creating cities that empower people rather than merely accommodate them. It poses the question of what ideals we decide to include into our urban forms and what kinds of lives our cities should sustain. This is already evident in Australia because to new frameworks that place a higher priority on social justice, walkability, liveability, and climate readiness. Instead of being influenced by market forces or short-term gain, future urban development must be motivated by deliberate outcomes—healthier lifestyles, closer community links, circular economies, and ecological balance. Abstract masterplans are giving way more and more to outcome-oriented urban policies in cities like Adelaide, Melbourne, and Brisbane, whose objectives are well-defined, locally relevant, and informed by the community.

Integrated urban design is the spatial translation of this intent. It creates a cohesive and flexible urban fabric by integrating housing, energy, water, mobility, landscape, and architecture. In Australia's context, this integration is crucial, not just for efficiency, but for identity and resilience. As a means of bringing big-picture thinking to the local level, precinct-based development concepts are becoming more and more popular nationwide. These projects, which range from Adelaide's Tonsley to Brisbane's Gabba Precinct and Melbourne's Southbank, show how design-led, multi-stakeholder cooperation can create excellent urban settings that are suited to location, people, and purpose. Integrated precincts are ecosystems of innovation, rooted in sustainability, culture, and community, rather than merely collections of buildings.

A key component of this future-oriented urbanity is still placemaking. By bridging the gap between the emotional and physical aspects of space, it guarantees that constructed spaces represent the identities, values, and memories of their occupants. Successful placemaking in Australia, a culturally varied country, necessitates considering local histories, Indigenous custodianship, and changing demographics. It entails developing locations that provide areas of purpose and gathering, encourage social cohesion, and attract involvement. When placemaking is done correctly, it humanizes development by transforming neighborhoods into communities and infrastructure into culture.

Looking ahead, Australian urban futures will depend on how well governance systems enable coordination, innovation, and civic trust (Osborne et al., 2021). Data and technology are essential, but they should be used to empower people rather than to monitor or exclude them. This is something I have reflected on when introducing the novel concept of "Advanced City" (Cheshmehzangi, 2025). With humility and forethought, the smart city ideal must transform into a wise city practice that listens, learns, and adjusts. Local governments, state authorities, businesses, academic institutions, and communities must all be involved in the planning process to build cities that are dynamic, equitable, and future-ready. In summary, our capacity to create spaces with a purpose, design systems with integrity, and exercise long-term governance will determine the future of Australian urbanity. If we are successful, our cities will not only adapt to problems but will also usher in a new era of urban existence based on compassion, ingenuity, and shared aspirations.

1.5.1 *Brisbane 2032: Catalyzing Urban Transformation Through Placemaking and Purpose-Led City Making*

I am not trying to be biased about Brisbane, but what is happening now is certainly a unique opportunity for the city, the region, and all our communities. The 2032 Olympic and Paralympic Games in Brisbane is a once-in-a-generation chance to change the city's urban fabric, strengthen its international reputation, and establish enduring community value. The 2032 Games are much more than just a major athletic event. Brisbane 2032, which is framed within the larger concepts of integrated urban design and purpose-led planning, aims to transform a city for future generations rather than just create for an occasion.

A strong dedication to placemaking lies at the heart of Brisbane's Olympic approach. Instead of constructing discrete "mega venues" (or white elephants), the Games are being utilized to promote inclusive precinct development, improve urban connections, and energize public areas. Human-scale design, community benefit, and cultural expression are the guiding concepts of projects like the renovation of the Gabba Stadium and the revitalization of the surrounding neighbourhoods. Green open spaces, active transportation, and mixed-use areas that benefit the city long after the closing ceremonies are overshadowed by these initiatives.

Another important component of the Games-led transformation is connectivity. Investment in sustainable mobility infrastructure, such as new train stations, active travel corridors, and the growth of the city's network of green bridges, is being accelerated by Brisbane 2032. The plan acknowledges that top-notch communities are bike-friendly, walkable, and transit-friendly. This objective is largely achieved by the Cross River Rail and Brisbane Metro projects, which include the Games infrastructure into a larger mobility program that promotes access fairness and long-term mode shift.

Identity creation is equally significant; a legacy that truly matters for us all in the city and beyond. Brisbane's subtropical climate, Indigenous heritage, and developing cultural identity are all purposefully reflected in the city's Olympic plans. A unique urban identity anchored in place is formed through the architectural language of new venues, the utilization of nature as infrastructure, and the centralization of local narratives in public space design. Brisbane's story—its river city character, its creative economy, and its changing prominence on the national and international stage—can be amplified through the Games.

Detailed strategies within Brisbane's Olympic planning emphasise legacy as a non-negotiable outcome (see Fig. 1.1). This includes:

- **Temporary to permanent infrastructure transitions**, ensuring facilities are scaled, modular, and adaptable.
- **Housing legacies**, such as the conversion of the Athletes' Village into long-term affordable and social housing.
- **Sustainability measures**, including carbon-neutral targets (with a revised narrative), passive design principles, and integration with blue-green infrastructure.

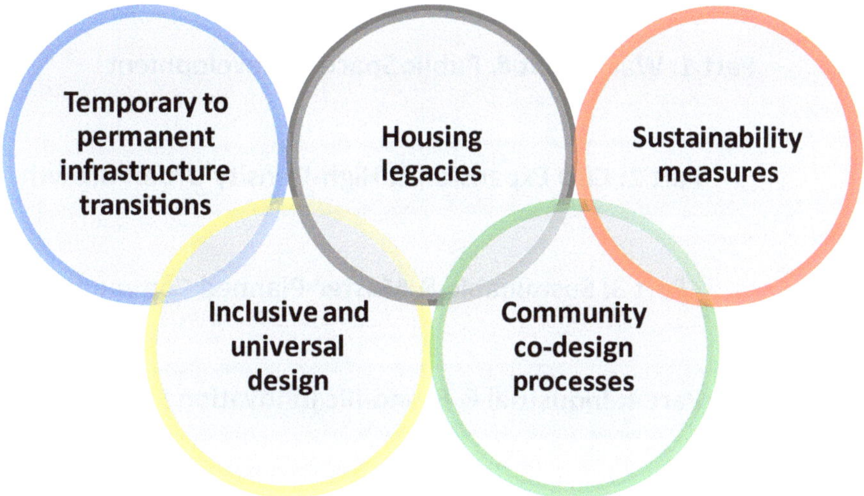

Fig. 1.1 Summart of detailed strategies within Brisbane's Olympic planning based on the five Olympics rings

- **Inclusive and universal design**, ensuring accessibility is embedded across all venues and public spaces.
- **Community co-design processes**, where local stakeholders are engaged early to shape developments that reflect their needs and aspirations.

Finally, Brisbane 2032 is a civic endeavor that reflects the changing story of Australian urbanity rather than merely being a test of design or logistics. It is a chance for the city to move away from a development-based city to a design-based city; a city that will be reshaped and redefined through transformations in infrastructure, public realm design, connections, and integrated urban design solutions in coming years. It centers a ten-year transformation around people, place, and purpose. Brisbane has the opportunity to set the standard for a new kind of event-led urbanism by utilizing the Games as a catalyst for placemaking, regeneration, and inclusive growth. This is one where legacy is planned for from the beginning and the city's pulse continues to beat strongly long after the world has gone on.

1.5.2 An Insight into the Book Structure and Its Five Overarching Themes

To wrap up this introduction chapter, the structure and thematic orientation of this book, which focuses on urbanity as well as the architecture, design, and city life features of Australian cities, should be quickly framed. The book is organized around five main topics, each of which represents a distinct way that cities are changing

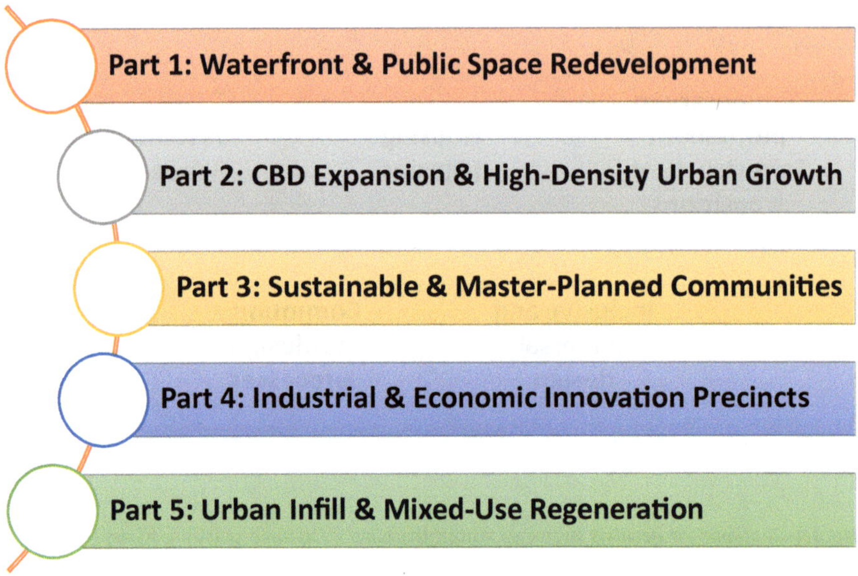

Fig. 1.2 Summary of our five themes in the book

in Australia. Each case study demonstrates a distinct aspect of how architecture, planning, and design interact with identity, sustainability, the economy, and public life; these issues are not discrete silos but rather are interwoven representations of Australian urbanity (Fig. 1.2).

Part 1: Waterfront & Public Space Redevelopment projects rethink the outskirts of Australian cities serve as the starting point for our investigation into redevelopment. These examples highlight how important public space is to urban identity and liability. Projects like Sydney's Barangaroo and Brisbane's South Bank demonstrate how once industrial or underutilized waterfronts may be developed into vibrant civic and cultural areas. This theme is furthered by Melbourne's Docklands, Adelaide's Festival Plaza, and Perth's Elizabeth Quay, which show how placemaking, connection, and landscape-led design are transforming the periphery of center cities into lively, inclusive spaces. These improvements are spatial tactics that give the city back to its citizens, not merely aesthetic fixes.

Part 2: CBD Expansion & High-Density Urban Growth emphasizes verticality, infrastructure, and connectedness. Densification and vertical integration of housing, transit, facilities, and workplaces are examples of urbanity in this context. Examples like Parramatta Square and Collins Square show the relationship between superior design and economic aspiration. The Brisbane Live Precinct and Civic Quarter in Canberra demonstrate how transport-led redevelopment can reshape city skylines and recast the rhythm of inner-city life, while Maroochydore City Centre offers a new central business district from the ground up that is intelligent, flexible, and digitally linked.

Part 3: Sustainable & Master-Planned Communities examines suburban growth/expansion using a new urban logic that is characterized by ecological responsiveness, community well-being, and long-term sustainability. These examples show how Australia has experimented with liveability models for the twenty-first century. The examples show how design and governance can create vibrant, future-ready neighbourhoods, from Ecco Ripley's green emphasis to Riverlea Park's balanced density, and from Yarrabilba's nature-based city structure to Nightcliff Foreshore's sensitive coastal restoration. In contrast, the Hills Showground Station Precinct advances a new suburban typology by combining density with transit.

Part 4: Industrial & Economic Innovation Precincts illustrates the transformation of once industrial districts and outlying areas into economic hubs. State-led initiatives to establish integrated employment centers are reflected in the Citiswich Business Park and Aerotropolis Western Sydney, while Lot Fourteen in Adelaide and Macquarie Point in Hobart demonstrate the adaptive reuse of key locations for research and innovation. A crucial test case for managing scale, sustainability, and connectivity while combining industrial heritage with mixed-use aspirations is Melbourne's Fisherman's Bend.

Part 5: Urban Infill & Mixed-Use Regeneration focuses on the city within the city and how adaptive reuse and infill construction are facilitating renewal without erasure, wraps out the thematic arc. Lessons on deliberate densification and heritage preservation can be learned from Main Drive Kew and Epping Renewal. Sydney's The Ribbon is a prime example of daring urban sewing, where public space, architecture, and infrastructure all come together. Multi-governmental cooperation and urban village concept are exemplified by the Darwin City Deal and Hobart's Glebe Hill Village, which revitalize existing locations.

These 25 case studies collectively make up the book's body under these five overarching themes. These themes are selected with a rationale; to understand diversity of urbanity in Australian cities as well as how they are manifested in the physical environments, spatial layouts, everyday lives, social experiences, and architectural interventions across all major urban hubs. They are a varied, location-specific representation of how Australia is rethinking its urban future rather than a single vision. As we proceed into the case chapters, this framework offers a lens for critical reflection as well as thematic clarity, painting a realistic yet ambitious picture of how Australian urbanity is changing, one project and one location at a time.

References

Boal IA (1998) The campus and the commons. Common Prop Resour Dig 46:1–4

Bose P (2024) Revisioning urban pulse. Urban Geogr 45(1):1–5

Buxton M (2014) The expanding urban fringe: impacts on peri-urban areas, Melbourne, Australia. In The security of water, food, energy and liveability of cities: challenges and opportunities for peri-urban futures, pp 55–70

Carson DA, Carson DB, Argent N (2022) Cities, hinterlands and disconnected urban-rural development: perspectives from sparsely populated areas. J Rural Stud 93:104–111

Cheshmehzangi A (2025) Advanced city: for the future of cities and opposing smart cities. Springer, Singapore

Cheshmehzangi A, Allam Z, Siew G (2024) Diversity in cities and addressing urban sustainability. In: Siew G, Allam Z, Cheshmehzangi A (eds) Diversity as catalyst: economic growth and urban resilience in global cityscapes. Springer, Singapore, pp 1–9

Cheshmehzangi A, He B, Zuo T (2025) *Strengthening health-resilience nexus through sustainable urban design in Brisbane, Australia.* In: Cheshmehzangi A, Zuo J, Sharifi A, Zhang R, Ziafati Bafarasat A, Zhao J (eds) Designing healthy cities: integrating climate-resilient urbanism for sustainable living. Springer, Singapore

McGaw J, Pieris A (2014) Assembling the centre: Architecture for indigenous cultures: Australia and beyond. Routledge, New York and London

Osborne C, Mayo L, Bussey M (2021) New frontiers in local government community engagement: towards transformative place-based futures. Futures 131:102768

Wasserman B, Sullivan PJ, Palermo G (2000) Ethics and the practice of architecture. Wiley, Hoboken, New Jersey

Part I
Waterfront and Public Space Redevelopment

Transforming waterfronts and public spaces into vibrant, accessible hubs that enhance liveability:

- **South Bank, Brisbane**—A premier cultural and recreational destination.
- **Barangaroo, Sydney**—A world-class waterfront precinct integrating business, residential, and green spaces.
- **Docklands, Melbourne**—A revitalized former docklands area with mixed-use developments.
- **Festival Plaza, Adelaide**—A cultural and entertainment hub enhancing Adelaide's city core.
- **Elizabeth Quay, Perth**—A waterfront redevelopment enhancing public spaces and connectivity.

Chapter 2
Living Edges with Purpose: The Pulse and Placemaking of South Bank, Brisbane

2.1 From Expo Legacy to Urban Vitality

Before South Bank became a celebrated cultural precinct and recreational anchor of Brisbane, the broader city itself was undergoing a period of urban maturation. Despite being perceived as a smaller metropolis than its larger eastern seaboard equivalents, Sydney and Melbourne, Brisbane has undergone a significant shift in recent decades. Brisbane, which was formerly known for its suburban sprawl, reliance on cars, and peaceful riverfront, is now a bustling city with a strong emphasis on sustainability, cultural expression, and people-centered design.

A deeper understanding of urban vitality—the notion that cities must provide more than just infrastructure and services, but also a vibrant and stimulating urban experience—is at the core of this change. By adopting this idea, Brisbane has worked to make the river, its most precious spatial resource, more active (Stimson and Taylor 1999). The river has become more and more integrated into the city's urban identity rather than being viewed as a border or barrier. Because of this change, the waterfront is now seen as a living spine for urban participation rather than an edge.

An important part of this development is the South Bank precinct (Cheshmehzangi et al. 2025). Being the most popular public area in Brisbane and a draw for both residents and visitors, it perfectly captures the city's increasing expertise in maintaining and fostering civic life. South Bank's ongoing importance and vitality comes from its connection with Brisbane's larger aspirations—to become a more inclusive, climate-responsive, culturally rich, and socially vibrant city—even though its change following Expo 88 is frequently cited as the spark.

In this sense, South Bank reflects Brisbane's metropolitan culture rather than being a singular accomplishment. Its strength lies in its capacity to unite disparate aspects, such as hardscapes and green spaces, play and art, commerce and reflection (Margalit 2019). It promotes hybridity and plurality by holding peaceful reading nooks, riverfront eating, festivals, protests, and contemplative walks all within walking distance

A. Cheshmehzangi, *Australia's Urbanity*, Urban Sustainability,
https://doi.org/10.1007/978-981-95-3881-2_2

of one another. It has assisted Brisbane in shedding its image as a city without a core and establishing one that is teeming with life, personality, and warmth.

Crucially, the subtropical climate of Brisbane has also influenced the development of South Bank. Instead of using templates from colder cities, the architecture and landscaping here are highly sensitive to seasonal variations, heat, shade, and humidity. Permeable surfaces, local vegetation, water features, and canopies are not merely decorative options; they are necessary for usability, comfort, and durability. South Bank is a prime example of designing with context and climate in mind.

Precincts like South Bank become more than just cultural assets as Brisbane grows, especially in preparation for major (or mega) international events like the 2032 Olympic and Paralympic Games. They serve as models for how the city may scale urbanity, liveability, and sustainability simultaneously. The insights gained from South Bank's convergence of programming, design, accessibility, and a purpose-driven vision are applicable much beyond its confines. This chapter explores how South Bank evolved into the center of Brisbane's urbanity and urban vision, rather than merely a revitalized riverbank.

2.1.1 A Place Reborn: The Legacy of Expo 88

One of Australia's most advanced examples of waterfront regeneration is South Bank in Brisbane, where placemaking, urban vibrancy, and cultural richness all come together. Its development from the 1988 World Expo site into a prestigious public area tells the tale of intentional urban design that supports accessibility, cultural inclusion, and a thriving sense of community in addition to urban rejuvenation. Expo 88's ruins have been recast as a catalyst for a completely new urban identity, not just reused.

It is impossible to overestimate Expo 88's influence on Brisbane's contemporary urban image. South Bank was a low-activity, industrial area on the south bank of the Brisbane River for many years. Following the Expo, civic impetus sparked political and design efforts to preserve the site's spatial and symbolic significance. A radical reimagining of how waterfront land could serve as a city's heart—a location where architecture, community, nature, and leisure all come together—was what came out of the redevelopment proposal (De Gruchy 2012). Several design-led interventions that put people before automobiles, culture before commerce, and publicity before privatization were started by the South Bank Corporation, which was created to supervise the precinct's development. As a result, the precinct strikes a balance between carefully planned architecture and unplanned activity.

With more than 14 million visitors a year, South Bank's 17-hectare precinct now offers a distinctive convergence of urban programming. As a cultural hub of the city, it is home to a variety of cultural organizations, such as the State Library of Queensland, the Queensland Art Gallery, the Gallery of Modern Art (GOMA), and the Queensland Performing Arts Centre (QPAC) (Bank 2010). But what gives South Bank its unmatched vibrancy is the unofficial public sphere, which lies behind the institutional layer. Every hour of the day is filled with diverse activity in the precinct, from the boulevards lined with arbour to the artificial Streets Beach, and from weekend markets to late-night riverbank parties.

South Bank's urban vitality has been planned, not just happened. With pedestrian-friendly networks that fluidly connect planned zones, natural ecologies, and commercial products, the site's architecture promotes porosity and permeability. Its public areas are thoughtfully designed to accommodate a variety of activities, such as yoga classes in the morning on the lawns, water play areas for families, movie screenings at night, and festivals that spread out over its open areas. Its ability to preserve a "city within a city" feel while catering to both international tourists and the larger metropolitan area is crucial to its success.

Crucially, in contrast to many mono-functional developments, the precinct does not withdraw at night. Its DNA is designed to activate at night. After dark, locals and visitors are still drawn to restaurants, bars, cultural events, and riverbank strolls thanks to well-considered lighting design, easy navigation, and a consistent feeling of security. South Bank is one of the few genuinely "round-the-clock" public precincts in Australia because of its 24-h movement, diversified uses, and cultural layering.

The social and economic statistics reflect this vitality. Due in part to the lifestyle benefits that come with being close to South Bank, property values in nearby communities like West End and Kangaroo Point have also increased. In addition, the precinct is vital to Brisbane's creative and tourist sectors, which help to establish the city as a top urban destination in the world. In contrast to many waterfront redevelopments across the world that have tended toward commercial exclusivity, its unfettered access to all public places guarantees that it stays inclusive. The spatial layout also exemplifies resilience. The architecture of South Bank was put through a rigorous stress test in the wake of the devastating 2011 floods in Brisbane. Its permeable surfaces, open spaces, and flood-conscious landscape design were crucial to its quick recovery and have subsequently influenced Brisbane's larger plan for climate variability adaptation and waterfront resilience.

South Bank represents not just a case of successful adaptive reuse, but a leading model in how waterfront urbanism can foster sustained vibrancy, inclusivity, and cultural magnetism. It offers a dynamic, changing precinct that embodies Brisbane's municipal ideals while avoiding the pitfalls of over-commercialization or sterile gentrification. It is a site where ordinary life meets spectacular, where ordinary walks are broken up by the unexpected, and where the city's riverbank turns becomes a binding, breathing edge of urban promise rather than a barrier.

The South Bank experience is a crucial point of reference for creating long-lasting, people-centered urban landscapes as the city grows and deals with issues of affordability, density, and climate change. It is a living example of what urbanity in the twenty-first century may look and feel like—authentic, plural, open, and alive—as well as a legacy from Expo 88.

2.2 The Sophistication of Urban Use at the Water's Edge

The smooth transition between South Bank and the riverside is the foundation of its spatial logic. South Bank views its riverbank as a shared cultural and environmental spine, in contrast to many waterfront developments that either commercialize or privatize the water's edge. Its urban design celebrates the river not as backdrop, but as an active co-author of experience—a living presence shaping how people move, dwell, interact, and feel. Compared to previous waterfront redevelopment concepts that frequently ignored the water or made it ornamental, this represents a significant change. The water's edge at South Bank is sacred, practical, and performative.

Maintaining openness and consistency along the perimeter is the precinct's defining feature. A wide, unobstructed promenade beside the river encourages strolling, resting, and observing. This design approach promotes municipal entitlement to the shoreline by being open rather than closed and porous rather than walled. Multiple sensory and social experiences are supported by the choice to incorporate soft edges, natural flora, and different terrains (such the water-play zones or the River Quay Green). It serves as a place for communication as well as movement—between the city and the river, between residents and visitors, and between people and nature.

Its urban fabric has been refined through the incorporation of human-scaled architecture, water-sensitive urban design (WSUD) techniques, and pedestrian promenades. Every spatial decision, whether it is the placement of trees for shade and stormwater collection, the use of permeable paving, or integrated bioswales, speaks to a multi-layered goal: to promote ecological health while facilitating a variety of urban activities. The result is a form of urban sophistication that avoids complexity and instead attends to the granular, i.e., comfort, beauty, usability, and adaptability.

With its curving steel structure covered with colourful bougainvillea, the renowned Arbour walkway has both practical and symbolic uses. As a functional spine, it directs pedestrian traffic from the performing arts district to the parklands, public beaches, and dining areas. It serves as a symbol of South Bank's identity as a precinct dedicated to fusing movement, environment, and art. Wayfinding becomes an experience thanks to this sculpture-like gesture that turns infrastructure into an invitation. A multi-sensory ecology—fragrances, seasonal color bursts, filtered light, and tactile materials that encourage interaction—enhances the urban life it animates in addition to the intelligent infrastructure that makes it possible.

South Bank's planned rhythm of urban use is among its most elegant features. South Bank pulses with a daily and seasonal choreography, in contrast to mono-functional precincts that only reach their pinnacle around weekend events or lunch. Joggers and yoga groups come in the early mornings; school groups, tourists, and gallery visitors arrive in the late mornings; families, students, and river-bank loungers take over the afternoons; and the cultural center—diners, outdoor entertainers, movie-goers, and strolling couples—is awakened in the evenings. South Bank's design layers space instead of dividing it up by user or purpose. It promotes intergenerational, intercultural, and intertemporal cohabitation without causing spatial friction.

Ecological sensitivity is inextricably linked to this level of design expertise. Nature is not considered décor at South Bank. Rather, it authentically and modestly references biophilic design ideas (Field 2006). The rainforest walkways use natural plants and habitat construction to create a subtropical microclimate of shade and retreat. The site's water-retaining wetlands and pocket gardens mitigate runoff and urban heat while softening the constructed edge. Green solutions that promote biodiversity and insulation are even found on the rooftops of surrounding buildings, including cafes and cultural centers.

The Streets Beach is arguably the most famous illustration of South Bank's playful interaction with its water's edge. A man-made lagoon with white sand beaches, it offers cooling facilities, kid-friendly entertainment, and a social environment of obvious inclusivity, all while challenging the traditional divisions between urban space and leisure. Few urban interventions have attempted, let alone succeeded, in democratizing access to water by reclaiming an Australian summer typology—beachgoing—into the center of the city.

The durability of this refinement is what sets it apart. Over thirty years after its founding, South Bank is still not just important but also a model. By refusing to base its identity on capital-driven improvements or immobile monuments, it has escaped obsolescence. By sponsoring temporary pavilions, renovating playgrounds, organizing new events, and allowing for everyday spontaneity, it has instead fostered adaptation. Its ability to adapt its spaces to shifting urban rhythms, rather than overdesigning, is what makes it successful.

To put it briefly, South Bank is a waterfront urbanism masterpiece that does not confuse vibrancy with visibility. Its ability to serve many without exclusion, its invitation to dwell rather than consume, and its interaction between lived culture and planned form are what make it so strong. It serves as a reminder that building along the water's edge doesn't have to be ostentatious to be remarkable. It just needs to be truthful, inclusive, and aware of the actual pulsations of urban life.

2.3 The Pulse of Urban Life: Culture, People, and Everyday Vibrancy

South Bank is a prime example of thick urbanism, as defined by urban theorists, which is characterized by a density of social interactions, overlapping functions, and layered meanings rather than a mere population. A constant pulse of city life, bright but flexible, curated yet natural, is created here by the blending of lived experience and spatial design. South Bank is more than just a place where things happen; it is an event in and of itself, one that is constantly changing and taking place in the lives of its residents on a daily basis.

The precinct's architecture prioritizes functionality above monumentality. The buildings, whether they are cultural icons like the State Library, the Gallery of Modern Art (GOMA), or the Queensland Performing Arts Centre (QPAC), do not take center stage in the city. Rather, they facilitate and frame interaction. These buildings serve as civic hubs, presenting discussions, performances, exhibitions, and educational opportunities that attract a wide range of audiences from all socioeconomic backgrounds. The non-institutional areas, such as the weekend craft fairs, the picnic-filled riverfront lawns, the community theater-animated amphitheaters, and the outdoor movie nights, are equally important.

Culture is not only displayed but also created in South Bank. From the massive installations that commemorate memory and identity to the transient pop-ups and interactive sculptures that encourage play and touch, public art is prominent and participatory here. The precinct serves as a living gallery, welcoming multicultural expressions and Indigenous stories that update the city's collective narrative on a regular basis. South Bank serves as the cultural backbone of annual festivals like the Brisbane Festival and the World Science Festival, while smaller, more regional events like dance classes, musical performances, or cultural rites blend in with the neighborhood's daily cadence.

South Bank stands out for its deliberate appreciation of informality. Without prescribing, the design actively encourages pause, play, and performance. An embankment of grass turns into a stage. A shady square becomes a place for meditation. Tai chi practitioners, skateboarders, street buskers, and language exchange groups all find a place to be and express themselves. These unplanned events are not disruptions; rather, they are essential to the vibrancy of cities. Here, urbanity is constantly co-created rather than imposed.

Spatial generosity serves as the foundation for this openness to different types of urban involvement. Terraced steps, shaded nodes, riverfront decks, and mini amphitheaters are some of the precinct's adaptable features that may be used for a variety of gathering and movement scales. Everything from private contemplation to massive public celebration is supported by these micro geographies. Crucially, South Bank does not strictly distinguish between work and play, official and informal, or weekday and weekend. It allows for coexistence rather than hierarchy by accommodating overlapping publics and multiple time signatures.

South Bank is also influenced by its closeness to educational and artistic hubs. Students, instructors, artists, and researchers visit the site every day from TAFE Queensland's South Bank campus and Griffith University's Queensland Conservatorium. Curiosity, experimentation, and a cultural vitality focused on the future are all brought to the area by this scholarly and creative presence. The end effect is a location that is both forward-thinking and introspective—a site of invention and legacy. More importantly, South Bank is an example of a democratic approach to cultural accessibility. Participation barriers are decreased by the fact that many performances and events are inexpensive or free. This openness is further reinforced by the precinct's physical logic, which includes broad promenades, translucent facades, and low thresholds. The message is unmistakable: everyone is welcome here, at whatever time of day or season.

South Bank provides an alternative at a time when many urban developments are criticized for prioritizing exclusivity and consumption. It exemplifies a coexistence-based urbanization in which people come to live, work, and belong rather than merely observe or spend. Brisbane's cultural agora is alive and public in the most profound sense, something that few metropolitan zones are able to maintain.

2.4 Placemaking as Purpose: Lessons from a Designed Identity

Perhaps South Bank's most lasting impact is its steadfast adherence to the notion that location is a lived experience rather than merely a question of spatial form. Each component—material, ecological, social, and symbolic—has been thoughtfully designed to foster a feeling of inclusion. This is a primary urban function rather than placemaking as an afterthought or aesthetic improvement. Users, not investors, are the precinct's main audience, and it embodies a place-led development strategy that values the life in between buildings over the grandeur of architecture.

South Bank has promoted an experiential richness based on common urban joy since its inception, from its initial design concepts to its most recent interventions. The selection of light levels in various public areas—brighter lighting along River Quay, softer tones beneath the Arbour—responds to intimacy and atmosphere in addition to safety regulations. The public furniture is purposefully diversified, including steps and ledges that serve as both seats and performance areas, open lawns for impromptu occupancy, and benches for resting. The colours and scents of plants gradually change throughout the year, reflecting seasonal rhythms and sensory layers. Even sound has been taken into account, with areas of peaceful retreat juxtaposed with areas that are lively due to buskers, water features, and neighbourhood gatherings.

This unified goal is supported by a governance style that has placed an emphasis on long-term stewardship rather than immediate profits. Through careful zoning, regular maintenance, and responsive design upgrades, the South Bank Corporation has struck a balance between civic character and commercial success. Examples of initiatives that welcome change rather than reject it include the pedestrianization of Grey Street, the thoughtful greening of hardscapes, and the installation of new infrastructure for community events. Importantly, however, individuality is never sacrificed for growth. Iterative, dialogic, and grounded in public value, this is adaptive urbanism.

South Bank's more apparent involvement with Indigenous culture is a crucial component of its changing identity (Ganis et al. 2014). The precinct has elevated Aboriginal and Torres Strait Islander presence in recent years through several commissions, consultations, and design gestures—not as an add-on to development, but as a fundamental story. Dual-language signs, ceremony areas, and Indigenous artworks incorporated into landscape architecture all allude to a more thorough reconciliation process. These components express cultural continuity and represent Brisbane's evolving urban awareness rather than being ornamental or symbolic. As a result, South Bank becomes a civic space for remembering, honouring, and reckoning in addition to entertainment and recreation.

In South Bank, placemaking is meant to be lived into rather than established. Because it maintains a cohesive character while remaining adaptable, the precinct prospers. This is a lesson that is all too frequently overlooked in waterfront reconstruction projects around the world, which may prioritize spectacle or status over public use and intimacy. Instead, by proving that genuine urban liveliness cannot be retrofitted, South Bank provides a counterbalance to commercial gentrification. It must arise from the diversity, routines, rituals, and goals of daily life.

Crucially, the inclusive culture of the precinct is the outcome of careful planning, tried-and-true programming, and user-centered design; it is not a byproduct. Its ability to support diversity without degenerating into incoherence is what makes it successful (Cheshmehzangi and Li 2020). South Bank has space for everyone—not figuratively, but practically—whether it's a university student seeking a peaceful study spot behind a tree, a toddler splashing in the water at Streets Beach, or an elderly person taking part in a communal ritual.

South Bank provides a different path as major cities continue to investigate the potential and danger of revitalized waterfronts: one in which architecture is a tool for public imagination rather than merely an outcome of economic reasoning. It teaches that identity in public space must be both anchored and open, that vibrancy is fostered rather than imposed, and that authenticity is developed over time. By doing this, South Bank transcends its status as a tourist attraction and a successful reconstruction project. It turns into a lived city, where life and design are inextricably linked rather than distinct.

2.5 Conclusion: Towards a Model of Living Urbanity

South Bank is neither static nor flawless. Its ability to maintain a solid placemaking foundation while remaining dynamic is what makes it successful. South Bank keeps evolving—not by reinvention, but through responsiveness—as Brisbane expands and evolves and as global cities face the problems of demographic complexity, climatic resilience, and cultural reconciliation. Its public sphere is still flexible yet cohesive, innovative but grounded in the values of fairness, accessibility, and enjoyment.

South Bank is a process—a long-term choreography of space, community, and meaning—rather than a finished design. It is a real example of how urbanity flourishes in open systems rather than finished goods, where design fosters interaction and policies maintain continuity without stifling spontaneity. South Bank provides a counterweight in a time when many waterfront redevelopments run the risk of becoming sterile or commercialized: a civic area that is incredibly local, immensely experiential, and widely accessible.

Its invitation to interact rather than consume might be its greatest asset. Its individuality is further enhanced by each path taken, bench used, and tale shared inside its areas. More than just a source of ideas, South Bank provides planners and urban designers with a dynamic model for how communities might promote living urbanity.

The everyday rituals of being together in space, time, and change are what give a city its vitality, not just spectacle.

South Bank maintains the rhythms of urban life without dictating them, from early morning runs along the river to evening performances beneath the stars, from leisurely interaction at Streets Beach to quiet times of contemplation under shaded canopies. It aims to host experience rather than regulate it. Because it never stopped listening to the city it serves, it is a place that belongs to everyone, which is maybe the greatest sign of success. South Bank serves as a landmark and a compass for Brisbane as it continues to determine its future, reminding us that great cities are not merely created but nurtured. Here, people, place, and purpose are all interwoven realities rather than distinct goals. South Bank transforms from a precinct into a living example of what inclusive, long-lasting, and human-centered urbanity may actually look like as a result of this interaction.

References

Bank S (2010) Queensland performing arts centre Cnr Grey and Melbourne Streets South Bank Queensland 4101 Australia. Available from: https://www.parliament.qld.gov.au/Work-of-the-Assembly/Tabled-Papers/docs/5309t1331/5309t1331.pdf

Cheshmehzangi A, Li HMA (2020) Innovation through urban diversity and achieving comprehensive sustainable urbanism from a community-oriented approach. Curr Urban Stud 8(2):222

Cheshmehzangi A, He B, Zuo T (2025) Strengthening health-resilience nexus through sustainable urban design in Brisbane, Australia. In: Cheshmehzangi A, Zuo J, Sharifi A, Zhang R, Ziafati Bafarasat A, Zhao J (eds) Designing healthy cities: integrating climate-resilient urbanism for sustainable living. Springer, Singapore

De Gruchy G (2012) Architecture and urban design in Brisbane, vol 1. Boolarong Press, Tingalpa

Field L (2006) Gardens of the mind: an exploration of the integral role of gardens in the lives of contemporary Australians by probing the construction of a public garden space and exploring visually an idiosyncratic artistic experience of the Flecker Botanic Gardens. Doctoral dissertation, James Cook University

Ganis M, Minnery J, Mateo-Babiano D (2014) The evolution of a masterplan: Brisbane's South Bank, 1991–2012. Urban Policy Res 32(4):499–518

Margalit H (2019) Australia: modern architectures in history. Reaktion Books, London

Stimson RJ, Taylor SP (1999) City profile: Brisbane. Cities 16(4):285–295

Chapter 3
Edge of Ambition in Urban Design: Barangaroo and the New Urban Imagination

3.1 From Working Harbour to Civic Statement: The Evolution of Barangaroo

For a long time, waterfronts have served as both a utilitarian and symbolic part of a city's identity. They have served as interchange points in the evolution of Australian cities, where infrastructure meets creativity, the economy meets ecological, and land meets the sea. The late twentieth and early twenty-first centuries saw a significant reconsideration of what waterfronts may be all over the world, with former port zones and industrial docks being repurposed as municipal spaces as well as redevelopment grounds. Sydney's western harbor redevelopment, Barangaroo, stands apart from the rest of the world due to its scope, ambition, and the stakes involved in its rebirth.

The Barangaroo site, which was primarily closed to the public and fashioned by logistical logic, was an industrial container terminal for more than a century. Its functional decline, like that of many other post-industrial waterfronts, produced both physical void and creative opportunity. Around the world, there was a change at the turn of the millennium: waterfronts were viewed as excellent sites for ecological restoration, mixed-use regeneration, and cultural expression rather than as the remains of a bygone industrial era. However, the Barangaroo project stands out because it was never just about rehabilitation or reuse (Pham 2017); rather, it was positioned as a national statement that exemplified what Australian urbanism could look like in the twenty-first century.

Situated on the outskirts of Sydney's central business district, next to Darling Harbour, and visible from the Harbour Bridge and Opera House, the site is both strategically and symbolically significant. The location required not only architectural change but also a rethinking of the city's spatial and social story because it served as a western entryway into the metropolis. It was supposed to be a new front door that would convey to locals and visitors alike sustainability, creativity, and civic identity (Cheshmehzangi 2020).

A. Cheshmehzangi, *Australia's Urbanity*, Urban Sustainability,
https://doi.org/10.1007/978-981-95-3881-2_3

But this transformation was never a blank-slate project. The growth and planning of Barangaroo have been influenced by political competition, public discussion, and conflicting ideas about what the area should stand for (Harris 2018). The site has served as both a civic forum and a design process, from the initial masterplanning phases to design modifications and public complaints. Its path has been shaped by issues with land use, heritage acknowledgment, Indigenous inclusion, ecological integrity, and the commercialization of public space. Architects, planners, corporations, governments, Traditional Custodians, and the general public have all engaged in a process of negotiation, visioning, and iterative communication that has resulted in the precinct that was created.

In this way, Barangaroo's development goes beyond merely erecting modern skyscrapers and waterfront parks in place of an outdated operational harbour (Logan 2015). It represents a more profound change in Australian urbanism, a move toward more performative, public, and disputed city-making. Barangaroo is a proposition rather than merely a precinct. A proposal on how cities might attract money without alienating the community, embrace density without losing their public character, and reinterpret history without erasing it. This birth story, of a site that started as a gateway for products and now serves as a doorway for ideas about the future of urban life in Australia, must be kept in mind while we examine Barangaroo's spatial features, cultural aspirations, and civic rhythms in the parts that follow.

3.1.1 Memory, Industry, and Opportunity

Politically, culturally, and economically, the location that is today known as Barangaroo has always held multiple layers of significance. It was a part of the Gadigal people's ancestral grounds in the Eora Nation long before it was turned into a logistics terminal or a redevelopment zone. Connected to profound knowledge systems of land and sea, this harbor edge sustained fishing, trade, and ceremony for thousands of years. A significant Cammeraygal woman from the early contact era is honoured with the name Barangaroo itself, a symbolic act of cultural recognition that still influences the precinct's identity and expectations.

New use and control regimes were brought about by the site's colonial and industrial layers. The region had undergone significant alteration by the nineteenth and twentieth centuries to make room for warehouses, shipping facilities, wharves, and eventually containerized freight terminals. Here was the embodiment of the maritime economy that shaped much of Sydney's ascent to prominence as a world port; it was utilitarian, functional, and mostly closed off to the general public. But by the 1970s and 1980s, the nature of international trade was changing. Inner-city working harbors like Barangaroo were abandoned as container shipping consolidated elsewhere and urban land values skyrocketed. This resulted in physical fragmentation and environmental degradation, but it also presented a unique opportunity: a sizable waterfront brownfield site on the outskirts of a bustling central business district.

The metamorphosis of Barangaroo is a perfect fit for the global story of waterfront restoration following industrialization. Sydney was forced to rethink what its western border could become as cities from Hamburg to New York, Melbourne to Copenhagen, attempted to transform abandoned harbors into vibrant mixed-use areas. However, Barangaroo's development was the focus of rigorous strategic planning and high-level political attention from the beginning, in contrast to many other sites that developed gradually or naturally. It had the potential to reaffirm Sydney's position as a global city of the twenty-first century by fusing commercial opportunity with ecological design, liveability with spectacle, and economic promise with the provision of valuable real estate.

Adaptation to climate change also came into play early. The location served as a test site for environmentally conscious design in the future because to its low height, exposure to the harbor, and close proximity to important commercial and transportation facilities. Thus, the redevelopment provided a platform to incorporate sustainability objectives, such as integrated water systems and carbon-neutral construction, while exhibiting resilient landscapes that could live with high-density residential and commercial activity. These goals were in line with the public's increasing desire for more than just wealth or beauty. In addition to having a striking appearance, Barangaroo was supposed to be a site that upheld moral principles by being socially inclusive, environmentally conscious, and mindful of its complex history.

Memory was still a controversial topic. The industrial past was not just erased but reinterpreted, often through material reuse, interpretive signage, and heritage consultation. However, some perceived the change as a rupture rather than a regeneration, particularly heritage advocates and maritime unions. In the meantime, Indigenous memory—which is older and more resilient—was progressively revived, not just in name but also via continuous engagement and artistic incorporation. As a result, the rebuilding of Barangaroo became a meeting point of ambition and memory, post-industrial desire and industrial legacy, design opportunity and symbolic weight. The complexity of the project—and its applicability to current urban discourse—lies in these tensions.

3.2 Verticality Meets the Water: Urban Sophistication at the Edge

One of the most difficult urban planning problems in modern Australian history is Barangaroo: how to balance the openness of the harbor with vertical density, and how to create a smooth transition between public accessibility and commercial ambition. The striking contrast between the wide promenades, open parklands, and a restored natural shoreline and the slender towers rising along the harbor's edge forms the basis of Barangaroo's spatial identity. It is a location where the city and the sea converge—not in opposition, but with a thoughtfully crafted invitation.

Barangaroo intentionally embraces verticality, in contrast to traditional waterfront constructions that withdraw from height out of respect for the horizon. A confident density is indicated by the skyscrapers of Barangaroo South, which is home to residential apartments, international financial institutions, and the Crown Sydney resort. Their sculptural, glass-dominant, and sleek construction is intended to provide permeability at ground level while making a statement on the skyline. Crucially, these buildings' podiums are intended to function as thresholds rather than barriers, with ground-floor programming, pedestrian lanes, and active frontages that bring life to the area on a human scale.

A rhythm between built form and open space is incorporated into the masterplan, which was first created by Rogers Stirk Harbour + Partners and later modified by regional design teams. Even as density rises, this rhythm is essential to preserving both physical and visual access to the water. From the expansive Wulugul Walk along the shoreline to the more intimate alleys, courtyards, and raised terraces nestled between buildings, the design preserves a hierarchy of public areas. Through scale, material, and movement, these transitions make sure that Barangaroo never feels closed off; rather, it provides a number of layered experiences that let people relate to place.

This capacity to reconcile disparate spatial logics—height and ground plane, built and unbuilt, formal and informal—within a single architectural language is what makes Barangaroo so sophisticated. Instead of overshadowing the public domain, the vertical features provide it the programmatic and economic density it needs to survive. The waterfront is curated, not abandoned. Continuous use during the day and into the night is supported by spacious promenades, covered seating, and integrated food and shopping. Public life is a co-author of the architecture, not an add-on.

Water-sensitive urban design (WSUD) is a key component of Barangaroo's design approach (Renouf et al. 2020). The precinct incorporates a number of ecologically responsive design elements in recognition of the site's ecological and climatic weaknesses, especially its susceptibility to storm surges and sea level rise. Permeable paving, bioswales, rainwater harvesting systems, and natural planting are all part of the precinct's spatial DNA and are not optional extras. One of Australia's first extensive carbon-neutral precincts, Barangaroo South, was created to show how high-rise urbanization and environmental sustainability may coexist together.

Another spatial dimension is added by Barangaroo Reserve, which is located at the site's northern end and purposefully returns to the terrain and ecological characteristics of the pre-colonial shoreline. This area, which was created by the landscape architectural firm PWP in partnership with local Indigenous communities, includes a headland park made up of over 10,000 hand-cut blocks of sandstone that were recycled from the site's initial excavation. In addition to restoring native planting systems, the reserve brings back natural forms that industry has long since destroyed. The park's mild incline towards the ocean, paired with winding walkways and terraces resembling an amphitheater, provides a reflective contrast to the towers' southward vertical push.

A complex urban choreography that embraces Sydney's global aspirations without sacrificing the tactile, democratic feel of the public domain is what unfolds across

Barangaroo. The horizontal and the vertical work together rather than against each other. An advanced urban condition is characterized by development that maintains its civic soul, density without detachment, and status without privatization. Therefore, Barangaroo's urbanity lies not just in its structure but also in the way it frames how cities can expand outward and higher without abandoning their residents.

3.3 The Pulse of the Precinct: Activity, Culture, and Public Life

Without a vibrant public life, Barangaroo's architectural aspirations and environmental innovations would be meaningless. Its street-level vitality—its ability to host, draw, and support daily life in both planned and unplanned forms—is what really makes Barangaroo a successful district, not its skyline. Barangaroo has established itself as a precinct where work, pleasure, movement, and culture all coexist in a daily rhythm that beats from dawn till dusk thanks to careful public realm architecture and well-planned programming.

The intentional planning and engagement of Barangaroo's continuous waterfront promenade, the Wulugul Walk, is a crucial component of its success (Rey and Harrison 2018). It serves as a communal area and a corridor for movement, running the entire length of the precinct and connecting to the larger Sydney Harbour Foreshore Walk. Here, a wide range of users and activities are created by the intersection of cyclists, joggers, tourists, office workers, and families. The promenade, which is lined with eateries, cafes, art exhibits, and covered lounging places, urges people to stay rather than just move. A continuous sense of connectedness to Sydney's character as a harbour city is fostered by the area's closeness to the water, framed views of the Harbour Bridge, and visual permeability between built form and open space.

Barangaroo's daytime pulse is heavily influenced by its status as a business hub. Global banking, technology, and law firms are housed in Barangaroo South's three commercial towers, which contribute to the area's strong professional weekday population. During lunch, these workers flood out into terraces and alleys, creating micro-public areas where informal socialization momentarily replaces official city life. However, Barangaroo was purposefully designed to run around the clock, in contrast to many business districts that close after hours. Evening life is maintained by the incorporation of casual dining establishments, waterfront bars, fine restaurants, and public performance spaces. With its hotel, casino, and dining options, Crown Sydney also contributes significantly to nighttime activity, but it also brings up crucial issues regarding equity in the allocation of resources and public access.

Barangaroo stands out for combining emergent culture with planned programming. The Barangaroo Delivery Authority and local partners collaborate to frequently stage public events such as night markets, art festivals, cultural performances, and sustainability expos. These gatherings strengthen the precinct's reputation as a hub for civic involvement while also attracting a varied range of tourists. In the meantime, the spaces between formal programs are used by passive activities, such as informal musical busking, peaceful reading in shady areas, or picnics on the grass. Both extremes of this spectrum are supported by the public realm's architecture. Wide promenades flow into tight, dynamic laneways; open lawns coexist with shady nooks; and private alcoves balance out large plazas.

One of the most important additions to this vibrancy is Barangaroo Reserve, the headland park at the northern end of the precinct. The Reserve is open to the public, non-commercial, and distinctly ecological in its design, in contrast to Barangaroo South's business district. It is popular with hikers, children, cultural groups, and fitness enthusiasts because of its tiered sandstone blocks, natural flora, and winding trails. It also functions as a venue for performances and events, including community meetings, seasonal celebrations, and Indigenous rites. Importantly, its shape and location provide broad vistas of the harbor, strengthening Sydney's emotional and visual bond with its waterfront. The Reserve restores not just a plot of land but also a feeling of inclusion in the city, particularly for people who are looking for a break from the bustle of the urban landscape.

The area is continuously activated by public art, which also gives the ordinary experience additional levels of significance. Leading Australian and international artists' creations are incorporated into Barangaroo's open and built spaces; they are not positioned as freestanding monuments but rather as part of the landscaping, light installations, and surfaces. The precinct's concentration on Indigenous art and storytelling, which is represented by plaques, sculptures, and explanatory signage, is especially noteworthy. These insertions serve as subdued provocations, asking users to consider the location's lengthy history and the voices that continue to influence it.

When it comes to accessibility and mobility, Barangaroo seamlessly combines a variety of transportation options, including buses, ferries, bicycles, pedestrians, and, most recently, the Sydney Metro. This improves the site's use on a daily basis and guarantees that it stays a part of the city, not a place isolated from it. Wide, shaded pathways and public elevators promote inclusive access for people of all ages and abilities, while entry points, signs, and thresholds are thoughtfully designed to guarantee legibility and simplicity of navigation.

This active public life does not, however, just happen. It is the result of long-term urban management, curation, and governance (Cheshmehzangi 2023). The Barangaroo Delivery Authority, which is currently a part of Infrastructure NSW, has taken the initiative to keep the area clean, safe, well-planned, and operational. However, public opinion also affects vibrancy, and in this regard, Barangaroo keeps changing. Some Sydneysiders applaud its change, while others criticize its alleged corporatization of public space and elitism. Striking a balance between openness and status, and authenticity and spectacle, will be the task going ahead.

Despite this, Barangaroo is now a high-performing urban zone where city life is not concealed or abstracted but rather unfolded in public—on benches, along walkways, in vistas and shadows, during performances and breaks. It is a space where design and action collide, and anyone is free to congregate, explore, work, eat, demonstrate, and perform. It provides an example of how purpose, programming, and people may be brought into real conversation to create urban vibrancy.

3.4 Designing Identity and Legacy: Placemaking in a Globalised Context

Barangaroo is a cultural endeavor that aims to integrate place identity into the logics of global capital, post-industrial city-making, and environmental consciousness. It is more than just a real estate success story or a historic precinct. It was always going to have symbolic significance because it was a waterfront development in the center of Sydney. However, Barangaroo's goal to combine global aesthetics with local belonging and to portray a forward-looking identity while navigating historical histories is what really sets it apart, not only its location or magnitude. The place-making challenge of Barangaroo in a globalized context is defined by this balancing act—between ambition and authenticity, between spectacle and delicacy.

Multiple, occasionally conflicting forces shaped the precinct's identity from the beginning. On the one side, Barangaroo had to function as a global precinct, including luxury facilities, multinational enterprises, and architectural features that would appeal to global consumers. However, it had to take into account Sydney's unique social, cultural, and ecological circumstances, including its harbour topography, Indigenous heritage, and changing civic ambitions. The end effect is a hybrid identity that combines elements of a cultural corridor, an ecological park, and a financial district. Although it is true, this hybridity is not always smooth. Barangaroo is a place where different expectations collide and history is actively created rather than inherited.

The intentional incorporation of Indigenous narratives and design elements throughout the precinct is one of the most obvious manifestations of this created past. The site's naming in honor of Barangaroo, a strong Cammeraygal lady from the early colonial era, represents a change in Sydney's public perception of whose histories are celebrated in the city's architecture. Dual-language signs, carefully chosen Indigenous artwork, interpretive displays, and ceremonial activities at Barangaroo Reserve are examples of tangible and symbolic gestures that go along with this naming, which is not tokenistic. These components show a rising dedication to place-based reconciliation in Australian urbanism, not just through acknowledgment but also through public visibility and design integration.

The precinct's efforts to integrate sustainability and history are equally significant. One of the first extensive carbon-neutral precincts in history, Barangaroo South was designed using a variety of climate-resilient, energy-efficient, and water-sensitive systems. This commitment to the environment supports the precinct's symbolic identity as a model for other cities in addition to its physical resilience. In this sense, Barangaroo promotes itself as a test site for potential urban futures rather than merely focusing on the past.

However, there are criticisms of this identity-building process. Many have questioned whether the public realm of the area is indeed open to the public or is covertly regulated, highlighting the conflict between commercial power and public space. Barangaroo Reserve's open culture is in stark contrast to Crown Sydney's closed atmosphere and privileged programming (Fullagar 2023). Furthermore, the prevalence of upscale dining and shopping has sparked worries about social justice and inclusivity, especially in a city that is struggling with segregation and poverty. The vulnerability of placemaking in high-investment circumstances is highlighted by these criticisms: identity must be continuously managed and defended rather than merely announced.

Notwithstanding these obstacles, Barangaroo shows that placemaking in an international metropolitan setting requires more than just branding or aesthetics. It calls for ongoing adaptation, collaborative visioning, and long-term governance. The Barangaroo Delivery Authority, which has since changed its name to Infrastructure NSW, has been instrumental in keeping the precinct's character dynamic and adaptable (Barns 2018). Barangaroo serves as a site for continuous civic negotiation through events, festivals, educational initiatives, and community collaborations.

Finally, Barangaroo's legacy might consist of multiple experiences and images rather than just one. It is a location that reflects the contrasts and goals of the modern metropolis: it is commercially strong, environmentally ambitious, historically aware, and culturally multifaceted. Its ability to continue to host, challenge, and change alongside its occupants will be used to determine its success rather than just architecture.

Barangaroo provides both a model and a warning as communities all across the world struggle with the challenges of waterfront renewal—where capital, climate, and culture converge. It reminds us that identity is not solely a result of design, but it also demonstrates what can happen when civic purpose is embraced by large-scale development. In response to the living dynamics of urban life, it needs to be earned, occupied, and continuously recreated.

References

Barns S (2018) Arrivals and departures: navigating an emotional landscape of belonging and displacement at Barangaroo in Sydney, Australia. Creative placemaking. Routledge, New York, pp 56–68

Cheshmehzangi A (2020) Reflecting on urban identities. In: Identity of cities and city of identities. Springer, Singapore, pp 133–178

Cheshmehzangi A (2023) City life experiences: lessons learnt for future urban regeneration strategies. In: Mapping urban regeneration: city life experiences in Yunnan, China. Springer Nature Singapore, Singapore, pp 239–246

Fullagar K (2023) Bennelong and Phillip: A history unravelled. Simon & Schuster, New York City and Sydney

Harris M (2018) Barangaroo: Machiavellian megaproject or erosion of intent? In: Urban regeneration in Australia. Routledge, New York, pp 111–134

Logan C (2015) The national trust and the Heritage of Sydney Harbour. Landsc Rev 16(1)

Pham K (2017) Examining the development of public space in foreshore Sydney: Barangaroo case study. Master's thesis, University of Technology Sydney (Australia)

Renouf MA, Kenway SJ, Bertram N, London G, Todorovic T, Sainsbury O, Nice K, Moravej M, Sochacka B (2020) Water sensitive outcomes for infill development: infill performance evaluation framework. Cooperative Research Centre for Water Sensitive Cities, Melbourne, Australia. CRC for Water Sensitive Cities, 3, 3

Rey J, Harrison N (2018) Sydney as an indigenous place: "Goanna walking" brings people together. AlterNative Int J Indig Peoples 14(1):81–89

Chapter 4
Reclaiming the Edge: Melbourne Docklands and the Search for Connected Urban Life

4.1 From Working Harbour to Urban Frontier: The Long Arc of Docklands

Docklands is a tale of aspiration, rebirth, and slow realization. For more than a century, Melbourne's industrial economy was driven by the Docklands precinct, which was once the city's maritime and logistical center. It was home to working-class labour, railway lines, cargo sheds, and shipping yards. Global changes in freight patterns and shipping technologies, however, made these areas obsolete by the late twentieth century. The area became underutilized and dilapidated as a result of the containerization trend and decentralized port infrastructure. All that was left was a huge stretch of beachfront land on the outside of the city center, secluded, broken, and teeming with untapped potential.

The Victorian Government launched a daring plan to develop Docklands into a new urban district in the 1990s after realizing the site's strategic importance (Shaw 2013). It was designed as a representation of the city's embrace of the knowledge economy, international investment, and contemporary urban lifestyles, in addition to being a geographical expansion of the Melbourne CBD. Docklands would be a civic statement—a flagship precinct that expressed Melbourne's future character as a connected, creative, and design-forward global city—rather than merely a rebuild.

But from the beginning, Docklands' size and complexity created inherent difficulties. The site, which was roughly 200 hectares in size and surrounded by arterial roads, the Yarra River, and railway infrastructure, was both geographically remote and alien to most Melburnians. Phased development was required due to its size and brownfield characteristics (Blyth 2002). Instead, the precinct developed through a number of sub-precincts, each created under public–private partnership models: Digital Harbour, NewQuay, Victoria Harbour, and Batman's Hill (Dovey 2013). Although this incrementalism made delivery more flexible, it also brought about long-term governance, public realm continuity, and design language fragmentation.

A. Cheshmehzangi, *Australia's Urbanity*, Urban Sustainability,
https://doi.org/10.1007/978-981-95-3881-2_4

VicUrban (previously Docklands Authority) was in charge of the development at first (Shaw 2013), and thereafter Places Victoria and Development Victoria took over. Private developers were in charge of providing public, commercial, and residential amenities, while these government organizations handled development agreements, infrastructure planning, and land release. Although the mixed-governance model was novel at the time (Harris 2025), it was criticized for putting market-driven results and quick land sales ahead of placemaking and integrated urbanization.

The promise of a mixed-use zone that would house 20,000 inhabitants, offer workspace for 60,000 workers, and draw millions of tourists each year through events, tourism, and waterfront recreation was at the heart of Docklands' initial ambition. However, it was more difficult to translate this vision into actual reality. Early stages of growth placed a strong emphasis on residential and commercial deliveries, frequently at the price of cultural programming, social infrastructure, and targeted public realm investment. As a result, the precinct felt unfinished in the middle of the 2000s; it was magnificent in terms of space but emotionally detached, and it was modern in terms of architecture but socially thin.

Public disapproval of what many referred to as Docklands' "placelessness" started to grow. The precinct lacks the vibrancy, sense of community, and multi-layered identity that characterize Melbourne's other urban neighbourhoods, despite its harbourfront location. Wind tunnels, deserted plazas, underutilized waterfronts, and a general inability to convey the vitality of urban life that characterizes the interior fabric of the city were among the criticisms levelled against it. Essentially, Docklands was charged with being a soulless new city.

Though legitimate, this criticism only provides a portion of the picture. Docklands has started a second arc of change in the last ten years, one that is more characterized by urban healing than by significant capital projects. The goal of placemaking activities has been to integrate the precinct into Melbourne's daily rhythms through temporary exhibits, event programming, public art projects, and upgrades to the area's infrastructure for bicyclists and pedestrians. The way Docklands is viewed and used has gradually changed as a result of investments in public space, including the renovation of Harbour Esplanade, community centers, and small parks.

Furthermore, Docklands is now viewed as a dynamic urban laboratory rather than merely a completed project. Newer planning techniques throughout Australia have been influenced by its mistakes, especially those related to monofunctional development, delayed activation, and an excessive reliance on private investment. At the same time, its advantages—close proximity to the central business district, direct waterfront access, and linkage to Southern Cross Station—continue to be potent pillars for future growth.

From this perspective, Docklands might be viewed as a learning precinct rather than a failed endeavor. It embodies the long-term strategy of post-industrial urban renewal, in which legacy is evaluated over time through integration, adaptation, and iteration rather than by instant recognition. It continues to evolve as a dynamic edge that Melbourne is still adjusting to, rather than as a frontier to be conquered in the city.

4.1.1 *Industrial Decline, Strategic Ambition, and the Politics of Renewal*

The historical and economic background of the location is presented in this part, covering its industrial shipping history as well as its downfall in the latter half of the twentieth century. It looks at the original investment and policy framework, early planning aspirations, and how the Docklands were positioned as a significant extension of Melbourne's central city. This subsection will look at how private sector involvement and government-led ambition interact to shape Docklands' future.

The present-day Docklands neighbourhood was one of Melbourne's most important industrial and maritime districts in the late nineteenth and early twentieth centuries. The region prospered as a center for manufacturing, storage, shipping, and rail freight because of its closeness to the Yarra River and easy connection to the growing Victorian rail system. The city's operational wharves, including Victoria Dock and Appleton Dock, served as the logistical hub of a labour-intensive, waterfront industry that was linked to international trade routes. Docklands was linked to municipal usefulness, marine employment, and industrial grit for generations of Melburnians.

But by the 1970s and 1980s, many inner-city docks had become ineffective and progressively outdated due to changes in global freight and port management, especially the rise of containerization. Docklands' operational importance decreased as shipping concentrated in bigger, outlying ports like Webb Dock and the more recent terminals at the Port of Melbourne. What was left was a vast area of waterfront land, around 200 hectares, that was mainly isolated from the rest of the city by infrastructure obstacles like arterial roads and train lines. It was also degraded and sitting idle. Similar post-industrial conditions observed in cities around Europe and North America were reflected in the deterioration of this landscape.

In order to unleash Docklands as a significant urban redevelopment site, the Victorian Government implemented a number of policy changes in the late 1980s and early 1990s after realizing the site's strategic urban potential. The Docklands Task Force was established in 1991 as the first significant step, and the Docklands Authority (later VicUrban) was established in 1997. These state-led initiatives marked a change in Melbourne's urban policy: public–private partnership models, zoning modifications, and infrastructure planning would all be used to deliberately coordinate extensive inner-city revitalization rather than relying solely on market forces.

The ambitious goal was to turn Docklands into a contemporary addition to Melbourne's core business area, a mixed-use waterfront center that would house public parks, entertainment venues, residential towers, multinational enterprises, and cultural attractions. Early planning documents' wording positioned Docklands as the next great urban frontier, one that would demonstrate Melbourne's ability to create daring urban designs and solidify its place as a global competitive city. The proposal was also presented in an economic framework, with the goals of boosting the construction industry, drawing in investment, and expanding the city's tax base.

But this goal was enmeshed in a very politicized environment. Large-scale land consolidation, rezoning, and a shift in public expectations were all necessary for the project. Like other massive revitalization initiatives throughout the world, Docklands also turned into a political tool, being utilized by succeeding administrations to show leadership and vision while also attracting criticism for privatization, open governance, and public benefit. The dominance of private developers, the dearth of affordable housing options, and the delayed delivery of promised public amenities were among the issues brought up by critics.

The chosen delivery model was purposefully hybrid. Land assembly, regulatory permissions, and basic infrastructure were the responsibilities of government agencies, and private-sector consortiums used expressions of interest to vie for development plots. Flexibility and market adaptability were made possible by this phased release approach, but it also made it more difficult to maintain architectural coherence and uniform public realm standards. Major developers including Lendlease, Mirvac, and MAB Corporation had established sizable footprints in Docklands by the early 2000s, helping to shape the changing cityscape with densely populated office, retail, and residential buildings.

The resulting district was mostly influenced by the logics of private investment, but it was also motivated by considerable public-sector ambition. Building was only one aspect of the problem; another was connecting—extending Melbourne's spatial and emotional fabric westward across previously unexplored territory. This was no easy task. The location was surrounded by freeways and railroads, which had long been the city's physical and mental boundaries. Infrastructure alone wouldn't be enough to integrate Docklands into Melbourne's daily life; cultural initiatives, story reinvention, and a persistent dedication to placemaking would be needed.

In this way, Docklands exemplifies the political complexities of late capitalist urban regeneration. It is a project born from strategic foresight yet shaped by market imperatives; a symbol of civic ambition, yet often contested for its authenticity and inclusivity. The results have been uneven, as the upcoming chapters will demonstrate, but the goals were unquestionably audacious. One of Australia's most important experiments with reclaiming industrial land and reimagining it for a different kind of urban future is still the Docklands project.

4.2 Shaping the Waterfront: Urban Form, Public Space, and Design Integration

The process of converting Docklands into a modern waterfront district has presented both logistical and design challenges. It took architectural clarity, infrastructural foresight, and a sophisticated grasp of scale to transform a vast, abandoned industrial area into a recognizable, dynamic urban setting. However, since its inception, the Docklands project has had difficulty striking a balance between the intimacy and complexity that characterize effective urban environments and gigantic ambition. Its development demonstrates not just a physical redesign of the waterfront but also a continuous endeavour to integrate various urban components into a cohesive whole.

The dispersed delivery mechanism of Docklands is one of its distinguishing characteristics. Docklands was constructed using a number of sub-precincts, including NewQuay, Digital Harbour, Batman's Hill, and Victoria Harbour, each of which was tendered to a different developer with their own architectural languages, schedules, and financial models, as opposed to being constructed from a single, cohesive vision. This led to spatial discontinuities throughout the precinct, but it also allowed for flexibility and market responsiveness. A lack of uniform character across several zones, sudden scale changes, and uneven public realm treatments were also pointed out by early commentators.

Despite these obstacles, VicUrban and later Development Victoria spearheaded a number of initiatives to bring the precinct together using broad design frameworks and urban design criteria. These frameworks sought to establish continuity among property parcels and along the waterfront by outlining standards for open space, pedestrian flow, connectivity, and architectural character. Wide walkways, bike lanes, and planting were given top priority in the design of streetscapes, and building heights and view corridors were controlled to maintain visual access to the lake.

The establishment of an uninterrupted waterfront promenade was a key component of the public realm strategy. The promenade, which stretches from Harbour Esplanade across Victoria Harbour and along NewQuay, was designed to serve as a public platform for movement, leisure, and entertainment as well as a connecting spine. In order to ease the boundary between land and water and encourage interaction, boardwalks, integrated seating, and public art pieces were added. However, people were frequently overwhelmed by the size of these coastal areas, especially in their early stages. Wide open plazas and promenades may seem deserted without enough programming, activation, or mixed-use frontages, particularly after hours.

Docklands' architectural style has been modern and frequently sculptural, especially in its office and residential skyscrapers. To create a recognizable character for the precinct, developers such as Lendlease and MAB Corporation hired well-known architects to create eye-catching structures. Examples of ecologically sensitive structures that contributed to the skyline's definition while embracing workplace innovation and sustainability concepts are Victoria Point, ANZ Centre, and the NAB headquarters. However, these buildings were frequently criticized for their lack of ground-level interaction, as under-programmed frontages, enormous foyers, and blank façades failed to foster a lively street life.

The integration of Docklands with the rest of Melbourne has been greatly aided by the transportation infrastructure. Accessibility and transportation legibility were enhanced by the extension of tram routes across the precinct, especially along Collins Street and Harbour Esplanade. Located on the eastern edge, Southern Cross Station serves as a significant multimodal hub by providing regional and city train connections. To improve cross-river and cross-precinct connections, pedestrian bridges like Seafarers Bridge and Jim Stynes Bridge were also erected. Early phases of development, however, did nothing to lessen the cutting effect of Wurundjeri Way and the

Flinders Street train corridor, which remained physical and psychological obstacles separating Docklands from the central business district.

The precinct's design approach has evolved over time to become more human-scaled. Ground floor activation, communal space inclusion, better lighting, shade supply, and support for small-scale retail have been the main priorities of more recent interventions. A shift toward people-first design is seen in initiatives such as the Docklands Community and Place Plan (2021), which is based on the knowledge that public life requires intimacy, narrative, and continuity in addition to infrastructure. Meanwhile, much-needed green space has been added by waterfront interventions like Docklands Park, Library at the Dock, and Ron Barassi Snr Park, which have helped to soften the built environment and foster acts of civic giving.

In terms of public space and urban form, the Docklands narrative is yet incomplete. While early phases leaned heavily toward scale, formality, and market-led outcomes, recent years have seen a recalibration toward urban coherence and community-centred design. The lessons are obvious: excellent waterfronts are shaped by the way areas are framed, utilized, and integrated into the daily lives of their people, not only by the design. The design challenge for Docklands as it develops is not to create the next famous skyscraper, but rather to create spaces that are inviting, embracing, and long-lasting.

4.3 Life at the Edge: Activation, Community, and Evolving Urban Pulse

The original plan for Melbourne's Docklands went beyond infrastructure and structures. It was always meant to be more than just a waterfront extension of the central business district; it wanted to develop into a thriving, socially active urban precinct that was teeming with activity, culture, and a strong sense of community. However, converting this concept into reality has proven to be one of Docklands' most difficult tasks. Perceptions of sterility, isolation, and underuse plagued the district for a large portion of its early history, which led to continuous efforts to create activation, promote social interaction, and strengthen the urban pulse.

Docklands' "emptiness" was criticized based on lived observation as well as architectural assessment (Rogers 2023). Large squares frequently seemed empty; ground-level stores had a hard time thriving; and the precinct's original residents, mostly office workers, helped to give it a strong presence during the week but left it largely deserted on the weekends and evenings. In the early stages of development, the absence of local services and integrated community infrastructure restricted citizen interaction and impeded the establishment of neighbourhood identity. To put it simply, Docklands lacked the human rhythm—the customs and casual encounters that give a city life.

Activation through cultural programming and event curation has been the focus of subsequent strategies in recognition of these limitations. A calendar of events has been put in place by the City of Melbourne, Development Victoria, and local stakeholders to encourage people to visit the area at various times of the day and year. These consist of food trucks, live music, art festivals, sport screenings, and seasonal pop-ups. Large-scale events like White Night, the Firelight Festival, and New Year's Eve fireworks have exploited Docklands' open shoreline and scenic setting to create spectacle and draw larger crowds.

The Library at the Dock, a community hub that debuted in 2014, is a particularly noteworthy addition to the precinct's civic life. In addition to providing books and digital resources, this first six-star green-star public facility in Melbourne also holds seminars, exhibitions, and community events. The transition of Docklands from a commercial experiment to a lived neighbourhood is both symbolic and practical. This shift has been strengthened by other programs that promote recreational and intergenerational use, like the community boating hub and Ron Barassi Snr Park.

With a carefully planned program of both temporary and permanent pieces intended to accentuate the landscape and provide moments of both visual and emotional engagement, public art has also contributed to the animation of Docklands. While temporary sculptures, light works, and street art have added texture to areas that previously felt too polished or generic, pieces like Reed Vessel and Silken Eyes provide interpretive responses to the site's maritime and Indigenous traditions.

The activation procedure has been slower at night. In contrast to the more established cultural districts of the inner city, Docklands continues to be perceived as having low nighttime liveliness, despite the fact that it is now home to an increasing number of clubs, restaurants, and entertainment venues. This feeling of fragmentation is exacerbated by the existence of large-format buildings and underutilized retail tenancies. To increase usage outside of regular business hours, precinct management teams have started experimenting with tactics like twilight markets, outdoor movies, and small-scale concerts.

Importantly, the population of Docklands is expanding in both quantity and variety. More residents, including young professionals, families, and international students, are choosing to live in the precinct, drawn by waterfront views, proximity to the CBD, and improved amenities. The patterns of community life become more ingrained as residential life does. To promote advocacy and connection, local associations, resident groups, and school outreach initiatives have been established. This bottom-up approach is essential because, although top-down activation tactics generate early traffic, grassroots involvement is what maintains long-term vibrancy.

Docklands project is still undergoing transition today. It is neither the completely developed precinct it aspires to be nor the failure that some early commentators characterized. It has a real, albeit erratic, urban pulse that is quieter during off-peak hours and loudest during scheduled activities. But there is promise in this flux. Docklands continues to learn, adapt, and react. Its planners and communities now realize that activation is a dynamic conversation between place and society rather than a single intervention. In this way, Docklands provides a lesson in the arduous process of bringing a place to life as well as an example of rejuvenation.

4.4 Learning from the Long Game: Identity, Adaptation, and Urban Futures

Over two decades after its first construction began, Docklands remains one of Australia's most important and closely watched urban renewal initiatives. Its tale is one of gradual adaptation rather than of sudden success or irreparable failure. As it develops, Docklands provides deep insights into the changing identity of post-industrial waterfronts, including their conception, use, and potential for transformation through bottom-up habitation and top-down design.

One important takeaway from Docklands is that placemaking, especially on this scale, cannot be hurried. The site's physical transformation happened quickly, but it has taken much longer for the social and cultural layers that are essential to creating a sense of place to develop. Early development frequently under-delivered on community infrastructure and human-scaled public places in favour of infrastructure, commercial tenancy, and high-density housing. The end effect was an environment that, despite being physically contemporary, found it difficult to emotionally connect with both locals and tourists. A deeper lack of lived narrative—spaces devoid of stories—was mirrored in the criticism of Docklands as being icy, desolate, or lacking in personality.

However, the precinct has started to re-establish its identity over time. Public opinion has changed as a result of the establishment of civic anchors like the Library at the Dock, funding for event programming, public art, and the gradual expansion of local community networks (Cheshmehzangi and Ornsby 2017). A dedication to co-designed futures as opposed to market-driven ones is demonstrated by the Docklands Community and Place Plan, which was created in consultation with locals and stakeholders. This change shows that maintaining long-term connections between location, people, and governance is just as important to urban rejuvenation as constructing new structures or infrastructure.

The significance of ecological interconnectedness is another important lesson. Hard landscapes and architectural form predominated at first, but Docklands has progressively added more green space, biophilic design ideas, and climate-resilient elements. The ongoing improvement of waterfront ecology and projects like Ron Barassi Snr Park demonstrate a growing understanding that environmental health is just as important to long-term urban vitality as economic or aesthetic worth. Docklands and other waterfront areas must take the lead in showing how dense urban environments can coexist with natural systems as cities deal with the challenges of heat, flooding, and biodiversity loss (Cheshmehzangi et al. 2023).

Docklands offers a sobering yet optimistic paradigm for urban futures. It demonstrates both the dangers of depending too much on private development to influence public life and the ability of long-term planning to steer things in the right direction. Despite its flaws, the precinct's governance architecture has made it possible to implement flexible tactics in response to criticism, shifting public preferences, and evolving civic expectations. Crucially, Docklands serves as a reminder that identity is fluid. It is negotiated, performative, and iterative. Both cities and a feeling of place are not created overnight.

Docklands adds to the larger conversation about waterfront transformation on a global scale. It symbolizes the potential and challenges of reclaiming post-industrial land for habitation, much like Toronto's Waterfront Innovation District, Hamburg's HafenCity, or London's Canary Wharf (Bullivant 2014). It has learned from its early mistakes. Its constant re-evaluation provides a realistic sense of hope—the recognition that even flawed endeavours can undergo significant change. Docklands' ongoing adaptation serves as a reminder that the life a place supports is a better indicator of an urban legacy than form or investment returns alone. Although its long-term goal is not yet complete, it is clearly headed in the right direction: toward more ecological awareness, inclusivity, and an urbanity that is felt rather than merely constructed.

References

Blyth I (2002) Reinventing Melbourne. The Building Economist, pp 4–9

Bullivant L (2014). From masterplanning to adaptive planning: understanding the contemporary tools and processes for civic urban order. Doctoral dissertation, London Metropolitan University

Cheshmehzangi A, He BJ, Sharifi A, Matzarakis A (2023) Climate change, cities, and the importance of cooling strategies, practices, and policies. In: Climate change and cooling cities. Springer Nature Singapore, Singapore, pp. 2–19.

Cheshmehzangi A, Ornsby P (2017) Augmented locality: the utilisation of urban screens in public places as new networks of the city. J Urban Regen Renew 11(2):194–211

Dovey K (2013) Dockscapes II–Precincts and projects. Fluid city. Routledge, New York, pp 157–181

Harris MS (2025) Docklands, Melbourne. Mixed-use megaprojects and the competitive city. Palgrave Macmillan, Singapore, pp 125–169

Rogers D (2023) Reclaiming the Blue Vague: activating urban waterways to create a sense of place. Doctoral dissertation, Swinburne

Shaw K (2013) Docklands dreamings: illusions of sustainability in the Melbourne docks redevelopment. Urban Stud 50(11):2158–2177

Chapter 5
Civic Thresholds: Festival Plaza and the Reimagining of Adelaide's Cultural Core

5.1 Between River and City: Festival Plaza's Context and Civic Role

One of Adelaide's most important locations in terms of both symbolism and space is Festival Plaza. The plaza serves as an essential link in the civic center of the city, sandwiched between the cultural splendour of North Terrace and the meandering margin of the River Torrens. Its location, right next to Parliament House, the Adelaide Railway Station, the Adelaide Festival Centre, and the Riverbank, places it at the intersection of formality and spontaneity, tradition and modern life, and institution and public (Llewellyn-Smith 2012). A deeper endeavour to rearticulate Adelaide's essential identity in a changing urban environment is evident in its recent redevelopment, which goes beyond a simple design intervention.

Historically, this area has always carried weight. The South Australian Museum, Art Gallery of South Australia, and State Library are all located on North Terrace, which has long served as the city's institutional and cultural thoroughfare (Vaughan 2021). On the other hand, the southern edge of the Torrens has been used as a background for performances as well as public parkland. But for a large portion of the late twentieth century, the area that is now Festival Plaza had trouble defining its urban function. The plaza itself served more as an ambiguous hardscape—a transportation corridor between the train station and Festival Centre—than as a real destination, even though it was surrounded by significant landmarks (Slynn and Conservatorium 2017). Its design, which dates to the 1970s, reflected a time of top-down civic monumentality and massive concrete geometry, frequently emphasizing formality over comfort and scale over intimacy.

It was becoming more and more clear by the early 2000s that this once-aspirational public area was not performing up to par. It did little to represent Adelaide's changing identity as a forward-thinking, culturally diverse, and internationally focused city, lacked activity, and did not encourage sustained involvement. More significantly, it

A. Cheshmehzangi, *Australia's Urbanity*, Urban Sustainability,
https://doi.org/10.1007/978-981-95-3881-2_5

created a psychological and physical divide between North Terrace and the Riverbank, which was a lost chance for event integration, pedestrian flow, and spatial narrative.

As part of a larger Riverbank Precinct Strategy—an ambitious, long-term plan by the South Australian Government to revitalize the western end of Adelaide's city center (Oakley 2021)—the drive for reconstruction gained traction. The plan acknowledged the River Torrens' unrealized potential as a geographical and metaphorical spine and sought to take advantage of its proximity to transportation hubs, cultural institutions, the Adelaide Oval, and biomedical innovation zones. As the focal point of this group, Festival Plaza was transformed into the municipal stage, a place where people might congregate, celebrate, demonstrate, perform, and just take in the city life.

Therefore, Festival Plaza's renovation was never solely a design endeavour. The site's ability to sustain both planned and unplanned applications, establish institutional gravity while expressing everyday liveliness, and function as a civic amplifier was crucial. The idea of reconnection was central to this goal. To do this, the river had to be reconnected with the city grid, people had to be reconnected with culture, and Adelaide's constructed legacy had to be reconnected with a modern idea of public life. It also entailed lowering the physical and psychological obstacles to space that had long limited the plaza's potential, enhanced pedestrian desire lines, and opening up views.

Festival Plaza's proximity to important vertical facilities further enhanced its strategic significance. Better accessibility to Parliament House and the Festival Centre increased civic permeability across institutional boundaries, while the integration of the Adelaide Railway Station through new gates and a public walkway allowed for more seamless north–south mobility through the site. In addition to enhancing circulation, these actions gave the plaza a democratic feel, implying that it belonged to everyone, not just politicians or ticket buyers.

In addition, the redevelopment project had to handle intricate financial, political, and design issues. The project, which was completed as a public–private partnership, included not only an improvement to the public realm but also the construction of a new commercial tower (the office building of Walker Corporation), the relocation of underground parking, and the development of a central plaza floor that integrated digital infrastructure, art, and lighting. The delicate task of striking a balance between these commercial imperatives and public benefit necessitated steadfast advocacy from design professionals, cultural groups, and the general public to maintain the importance of the public experience.

Festival Plaza is now an urban statement as well as a revitalized area. It declares that even historically static or inactive spaces may be transformed into vibrant, welcoming, and contextually aware settings. It no longer caters to a specific purpose or group of people as a civic platform. It functions across programs, from public protest to private play, and across temporalities, including weekend get-togethers, afternoon performances, morning commutes, and nighttime illumination. Festival Plaza, which is situated between the city and the river, serves as a reminder that the most significant urban areas are those that accept their complexity, work through their history, and allow for future interpretations.

5.1.1 *Historical Layers and Strategic Intent*

The character of Festival Plaza is intricately linked to the complex civic, cultural, and political topography of Adelaide. The location, which is tucked away between North Terrace and the River Torrens, has long been a literal and metaphorical pivot in the urban framework of the city. It is essential to trace the historical and physical settings that have created its function—and its limitations—as a public domain in order to comprehend its metamorphosis into a revitalized public plaza.

Part of Adelaide's "cultural boulevard" along North Terrace, the area is home to some of the most significant civic institutions in the state, such as the University of Adelaide, Parliament House, the South Australian Museum, the State Library, and the Art Gallery of South Australia (Rofe 2023). Founded on the principles of civic order, education, and cultural refinement, these institutions are representative of Adelaide's Enlightenment-era planning philosophy (Hay et al. 2004). The idea of a formal forecourt for the Adelaide Festival Centre arose in the middle of the twentieth century as modernist planning principles started to impact public space design. This was part of the city's goal to become a cultural capital.

As Australia's first multipurpose arts arena, the Adelaide Festival Centre opened its doors in 1973, solidifying the region's reputation as the creative hub of the city. The adjoining plaza was originally intended to be a huge modernist municipal square, surrounded by brutalist buildings, angular concrete shapes, and enormous steps. However, this area found it difficult to stay relevant throughout time. It still had cultural meaning, but it served no practical purpose. The spatial design often felt uninviting—too stark, too vast, and too disconnected from the pedestrian life of the surrounding city. It was more frequently utilized as a corridor than a destination, even though it was close to the river and North Terrace.

Public expectations changed in the next decades as a result of shifting urban life patterns, the emergence of more participatory public space models, and the increasing significance of placemaking. What was before considered "civic grandeur" started to seem archaic. Specifically, the plaza's inability to accommodate the informal, inclusive urban life that characterizes successful modern civic spaces was hampered by its lack of human-scaled architectural elements, sparse seating and shading, and lack of integrated programming. The location no longer accurately reflected the lively cultural image Adelaide aimed to create, even though it was still used for events and as a gateway to the Festival Centre.

The South Australian Government realized in the early 2010s that a significant intervention was required, not just to revitalize Festival Plaza but also to make a significant contribution to larger riverside rehabilitation initiatives. The Adelaide Oval, the Convention Centre, biomedical precincts, and improved parklands are all part of this area, which was being redeveloped as a unified cultural, athletic, and innovative corridor (Blakely et al. 2019). Festival Plaza, both physically and metaphorically, was at the center of this goal. Its transformation was a crucial urban plan to connect the city, river, and people, not a stand-alone design project.

The government's larger objectives for Adelaide's city center revitalisation, a multifaceted endeavour to revitalize the central city, encourage cultural tourism, promote walkability, and strengthen public transport connectivity, were also in line with the redevelopment project. The project's significance in repairing the city center's broken boundaries was further enhanced by Festival Plaza's proximity to the Adelaide Railway Station and the planned addition of new pedestrian linkages between King William Road, the Riverbank Footbridge, and Parliament House.

In terms of strategy, the redevelopment was treated as a public–private collaboration, integrating the construction of a new commercial office tower and underground parking created by Walker Corporation with the supply of improved public space. Public access, commercial encroachment, and the visual dominance of new built form were the main areas of concern for this approach. But it also made it possible to invest in a first-rate public space that was created by ARM Architecture and TCL (Taylor Cullity Lethlean) with the goal of balancing civic formality with human-centered design.

Thus, the change in Adelaide's urban thinking from static monuments to dynamic, inclusive, and performance spaces is reflected in the remodelling of Festival Plaza (Milne 2004). It tells the tale of how a legacy civic forecourt, which was formerly characterized by institutional distance and heavy form, is being redesigned to accommodate the rhythms of modern public life, bringing people closer to the arts, to the government, and to one another in the most prominent and symbolic area of the city.

5.2 From Hardscape to Human Scale: Designing Public Realm and Movement

An example of how public space can change from an austere municipal void to a multi-layered, inclusive, and navigable public realm is the redevelopment of Adelaide's Festival Plaza. Concrete terraces, expansive stairs, and sharp geometries characterized the original site, which was influenced by modernist design ideas of the 1970s. Although the purpose of these forms was to communicate architectural difference and institutional strength, they also created a plaza that was unconnected to the rest of the city and unsuitable for informal or casual purposes. In response, the new Festival Plaza takes a completely different tack, emphasizing elements of human-centered urbanism such as pedestrian movement, landscape integration, spatial openness, and material warmth.

Redefining the plaza as a connecting surface; an open, flexible plane that can accommodate both regular traffic and special events, rather than as an item or monument was one of the main goals of the renovation. This method necessitated reconsidering how building form, mobility, and landscape relate to one another. Previously, abrupt level shifts and a lack of direct permeability disrupted the site's gently sloping topography between North Terrace and the River Torrens. A more intuitive pedestrian experience was made possible by the redesign's gradual ramps, landscape terraces, and visual continuity, which helped to express and ease these changes. The result is a plaza that feels accessible from all sides, no longer a space to be bypassed, but one to be entered, lingered in, and moved through with ease.

Under the direction of Taylor Cullity Lethlean (TCL), landscape architecture was essential in bridging the gap between human-scale use and civic institutions. A carefully chosen landscape palette that incorporates native plants, shade trees, and precisely calibrated lighting softens hardscape components like granite paving, water features, and integrated seating. The vegetation serves as spatial infrastructure, marking unofficial gathering places, improving thermal comfort, and producing a changing sensory experience depending on the season and time of day. It is not attractive. By linking the architectural form of the city with the ecosystem of the riverbank, these landscape insertions also make reference to the neighbouring Karrawirra Parri (River Torrens).

The flexible open plaza floor at the center of the design is meant to be used for a variety of purposes, including public performances, civic gatherings, leisure play, and passive enjoyment. This multifunctionality is accomplished by restraint rather than overprogramming. The surface is purposefully open and spacious, providing room for movement, and is surrounded by digital screens, art pieces, and lighting columns that can be turned on or off based on the situation. This illustrates a recognition that outstanding public areas need to accommodate both nuance and spectacle; they need to function for both five and five thousand people.

Changing the space's environment also required careful consideration of the materials used. The new design makes use of lighter stone, timber accents, and reflecting surfaces that capture and react to Adelaide's abundant natural light, in contrast to the original plaza's heavy reliance on exposed concrete and deep shadows. Underfoot, the paving pattern adds texture and rhythm, drawing inspiration from regional geology and cultural themes. In addition to being aesthetically pleasing, these tactile components facilitate movement, deconstruct scale, and provide quiet intimate moments in an otherwise open space.

Enhancing interaction with other structures and establishments, particularly the Adelaide Festival Centre and Adelaide Railway Station, was another goal of the renovation. In addition to improving circulation and readability, new entrances, ramps, and pedestrian corridors also lessen the perceived institutional distance. The lower levels of the Festival Centre are now more physically and aesthetically connected to the plaza, allowing for unplanned interaction between cultural events and onlookers. In a similar vein, the upgraded interface with the train station turns what was formerly a transitional area into a destination, signalling the start of an urban experience rather than just the conclusion of a commute.

The plaza's success depends on this reactivation of edge conditions. In the past, uninviting thresholds, blank walls, or vehicle infrastructure dominated the site's borders. The new layout, on the other hand, views the plaza's borders as lively edges where people may dine, congregate, and observe. In order to promote safety and social engagement, new commercial tenancies, seating areas, and shaded platforms provide chances for informal occupation and visual surveillance.

Crucially, the design successfully reconciles formal and casual applications. The Festival Center's artistic significance and Parliament House's grandeur are not lessened; rather, they are placed in conversation with commonplace usage, such as office workers eating lunch on stepped platforms, students relaxing beneath trees, or kids playing in fountains. A mature urban sensibility is shown in the coexistence of public spontaneity and governmental authority, which maintains municipal identity through daily, diversified, and democratic presence rather than just ceremonial performance.

Ultimately, the renovated Festival Plaza is a prime example of how well-considered public realm architecture can humanize expansive public areas by establishing invitation, intelligibility, and permeability. Instead of imposing its will on its people, it adopts an architecture of involvement, moving away from the ostentatious but meaningless gesture. By doing this, it affirms the function of public squares as lived, felt, and changing aspects of the urban everyday rather than just as formal platforms.

5.3 Culture in Motion: Activation, Programming, and Everyday Life

Festival Plaza is a dynamic stage where Adelaide's cultural life takes place on many scales, in a variety of eras, and for a wide range of audiences. It is more than just an open area surrounded by building. The plaza was always meant to be a site of performance and gathering, as its name implies, but in its new and redesigned shape, it becomes something more democratic and adaptable: a place where community expression, everyday uses, unofficial traditions, and carefully planned cultural events all coexist in ongoing conversation. The public realm's function as an engaged and responsive civic platform is further supported by this state of "culture in motion", which encapsulates the changing urban pulse of central Adelaide.

The redevelopment's capacity to accommodate both planned and unplanned activities is a major factor in its success. At one extreme, the plaza is the site of massive gatherings that attract thousands of people, such as civic ceremonies, public broadcasts, arts festivals, and community celebrations. Live music performances, Adelaide Festival events, and light installations that emanate from the nearby Festival Centre are all supported by it. Because of the main plaza floor's integrated lighting, electricity, and sound systems, event planners can alter the space to accommodate a range of public events and performances. Programmable LED lighting that is integrated into the infrastructure and ground plane at night turns the area into a vibrant, changing canvas that provides drama, atmosphere, and a feeling of occasion.

The space's ability to promote daily vitality is equally significant. Festival Plaza stays vibrant outside of festival times thanks to unofficial and passive uses, such as people getting together for coffee, kids playing by the water features, workers walking from adjacent companies for lunch, and students relaxing in the shaded seating areas. Even though they are brief, these moments add up to a living rhythm that makes the plaza lively even outside of its main events. The design purposefully blurs the lines between public life and performance; for example, a busker's music melds with the background noise of the city, a digital art projection draws in onlookers, and a noon picnic is held in front of a political demonstration. The characteristic of a successful urban area is the layered coexistence of formal and informal culture.

The plaza's experience quality is further improved by the incorporation of lights and digital art (Fazakerley 2012). The design incorporates technology into the space's infrastructure rather of treating it as an add-on, giving community organizations, artists, and institutions a platform to exchange real-time information, cultural content, and visual tales. By reacting to events, seasons, and even the time of day, these components make sure that the plaza is dynamic and reinforces a feeling of change and rhythm. In addition to being aesthetically pleasing, lighting solutions increase safety, attract foot traffic, and prolong the space's usability into the evening.

The growing prominence of Indigenous design narratives and expression is essential to the plaza's cultural character. The plaza is situated on Kaurna soil, and an attempt has been made to represent this tradition in both tangible and intangible ways. Indigenous culture is acknowledged and celebrated through the use of language, patterns, and storytelling motifs in ground treatments, artwork, and interpretative components. These are ingrained cultural layers rather than tokenistic additions; they were frequently co-designed with Kaurna artists and cultural guardians. These actions transform the area into a location of cultural continuity, acknowledgment, and healing in addition to being a modern civic platform.

The way the plaza operates as a civic area has been greatly influenced by public involvement. Feedback from locals, business owners, cultural institutions, and community organizations shaped access plans, programming, and amenity offerings during the reconstruction. Through locally driven projects, community pop-ups, and co-curated events, the plaza's operation has maintained this participatory approach. In this context, activation is something that is developed internally rather than something that is imposed; it is an urbanism of cooperation.

With a tempo that varies throughout the week and year, Festival Plaza now functions as a sort of urban heartbeat. The area depicts a city that is learning to reconcile civic heritage with modern living, from the peaceful commuter morning routines to the vibrant festival nights. It is more than just a setting for artistic expression; it is an integral element of the performance, changing its meaning and mood to fit the city it serves.

5.4 Placemaking as Reconnection: Lessons from Adelaide's Central Urban Heart

One of Adelaide's most significant attempts to re-center public life through a purpose-led design approach is the renovation of Festival Plaza. In addition to physically transforming a once-underutilized area, it also reframes municipal identity at a pivotal point where heritage, modern urbanism, and forward-thinking goals meet. The plaza, which is situated at the intersection of culture, politics, transportation, and daily mobility, has developed into a new type of civic commons that seeks to bring people back to the center of urban life.

Festival Plaza is fundamentally a connectivity effort. reestablishing ties between the North Terrace cultural axis and the River Torrens, between official institutions and unofficial public expression, and between Adelaide's complex past and vibrant present. Flow, porosity, and spatial invitation have replaced the hard-edged transitions, fragmented levels, and underutilized surfaces that were formerly present. The plaza's ability to make public space accessible not only physically but also emotionally and symbolically is what makes it so successful.

However, there have been certain restrictions on the project's development. A component of the larger public–private redevelopment agreement, the commercial tower next to the plaza has come under fire for its size and possible violation of public realm values. The long-term equilibrium between corporate presence and civic openness is still up for debate. Furthermore, even if the site's spatial architecture is admirable, it will require consistent programming, upkeep, and inclusive governance that prioritizes community voices in order to reach its full potential.

Festival Plaza provides important insights for mid-sized cities looking to revitalize their urban centers without sacrificing authenticity, despite these conflicts. It demonstrates that placemaking is about establishing the circumstances for public life to flourish across social, cultural, and temporal layers rather than just about aesthetics or architectural status. Adelaide's strategy, which combines event programming, landscape integration, and cultural clustering, shows how even small cities can use design to foster vitality, community, and pride.

The project also emphasizes how crucial it is to see public space as a dynamic ecosystem rather than a fixed background. According to this perspective, places like Festival Plaza are influenced by and part of larger networks of meaning, mobility, and memory rather than being in a vacuum. In addition to form, their success hinges on flexibility, cultural resonance, and a willingness to be reinterpreted over time.

Festival Plaza emphasizes the importance of civic symbolism in modern urbanism in perhaps the most potent way. The plaza serves as a reminder that everyone should have access to the heart of a city in a time when public space is becoming more and more restricted or commercialized. It represents a revitalized faith in Adelaide's urban future, one that respects its history while embracing a more open, participatory urban lifestyle. Festival Plaza is therefore a process rather than just a location. A dynamic structure of individuals, narratives, and exchanges. a threshold area where people may see, hear, and share the city's pulse. It provides a sensible example for cities around the world that aim to mend the rifts in their urban fabric with kindness, creativity, and consideration (Cheshmehzangi and Munday 2022). Its lessons extend well beyond Adelaide.

References

Blakely EJ, Hu R, Blakely EJ, Hu R (2019) Pursuing innovation in Australian cities. In: Crafting innovative places for Australia's knowledge economy, pp 155–190

Cheshmehzangi A, Munday R (2022) From global to local: the case of migration and urban identity for regenerative city transformations. J Urban Regen Renew 15(3):308–324

Fazakerley R (2012) Art and the urban plaza: from landscape to environment. In: Urban transformations booms, busts and other catastrophes, pp 97–114

Hay I, Hughes A, Tutton M (2004) Monuments, memory and marginalisation in Adelaide's Prince Henry Gardens. Geografiska Annaler: Series B Hum Geogr 86(3):201–216

Llewellyn-Smith M (2012) Behind the scenes: the politics of planning Adelaide. University of Adelaide Press, Adelaide, p 400

Milne G (2004) Theatre Australia (un) limited: Australian theatre since the 1950s, No. 10. Rodopi.

Oakley S (2021) Waterfront regeneration in Australia: local responses to global trends in reimagining disused city docklands. Geogr Res 59(3):394–406

Rofe MW (2023) Memorial landscapes, recognition, and marginalisation: a critical assessment of Adelaide's 'cultural heart.' Landsc Res 48(5):632–646

Slynn D, Conservatorium E (2017) Music, place and public space: street music in Rundle Mall, South Australia. Doctoral dissertation, Master's thesis, Elder Conservatorium of Music, Adelaide, Australia. Retrieved from https://digital.library.adelaide.edu.au/dspace/handle/2440/118608

Vaughan G (2021) Public art museums in Australia: a brief history. In: A companion to Australian art, pp 51–78

Chapter 6
Reuniting City and River: Elizabeth Quay and Perth's Civic Reawakening

6.1 Divided by Design: The Historical Disconnection of Perth and the River

Elizabeth Quay, a project that aims to reconnect a city with its natural and cultural edge, is a significant urban recalibration for Perth (Cooper 2015). Despite the Swan River's ecological and aesthetic significance in the city's geography, central Perth was oddly distant from it for a large portion of the twentieth century. This disparity in space and perception was not accidental; rather, it was the result of several planning choices, the predominance of infrastructure, and a built form that prioritized transportation and efficiency over public interaction with the water's edge. Perth was similar to many other cities in Australia and abroad in this regard, since they had abandoned their waterfronts and viewed them as places of utility rather than public life.

Perth's growth was historically largely influenced by the Swan River. The river, which the Whadjuk Noongar call Derbarl Yerrigan, has long held cultural, spiritual, and utilitarian significance (Critchley 2022). For thousands of years, it influenced Indigenous myths, seasonal cycles, and paths. Jetties, warehouses, and port facilities were built along the river's northern banks as a result of European colonization in the nineteenth century, which turned the river into a transportation and trading route. Roads, parking lots, and a network of traffic infrastructure, notably Riverside Drive, replaced these purposes as the city modernized over the twentieth century, essentially separating the central business district from the river. The Swan appeared but was out of reach; it was there in picture but mostly not there in reality.

The loss was civic as well as spatial. The center of Perth, which was centered on shopping centers, government buildings, and remote business districts, had grown more reclusive by the middle of the twentieth century. Vehicle traffic, unused public lawns, and bare edges dominated what ought to have been the city's most valued civic threshold—the intersection of land and river. The terrain that had given Perth its purpose was separated from civic identity, public culture (Andrews et al. 2024), and ecological consciousness.

A. Cheshmehzangi, *Australia's Urbanity*, Urban Sustainability,
https://doi.org/10.1007/978-981-95-3881-2_6

The idea of Elizabeth Quay surfaced as a daring counter-narrative against this background. Elizabeth Quay was framed as a symbolic and practical gesture when it was announced in 2011 as a flagship project of the Western Australian Government's "Directions 2031" urban growth strategy. The goals were to reestablish the waterfront as a hub for public gathering, cultural production, and economic vitality, as well as to reconnect Perth with the river and reframe the city's global image. It was an attempt to change the city's orientation, to expand its civic footprint toward the lake and establish a new kind of urban edge, rather than just a beautifying project.

The choice to remove a portion of Riverside Drive in order to make room for a new inlet that would reintegrate the river into the city was important to the project. This action was intellectual rather than merely infrastructural. Instead of just passing it quickly, it allowed people to walk, cycle, and hang out along the edge, restoring a sense of porosity between the city and the ocean. A move toward human-centered design was signalled by the addition of a public bridge, open municipal spaces, a pedestrian promenade, and integrated transportation links.

Crucially, the project had to strike a balance between the past and the future. Any action would have symbolic significance because Elizabeth Quay is close to famous locations including the Supreme Court Gardens, the Bell Tower, and the base of St George's Terrace. The quay must both herald a new age of urban inventiveness and tell the tale of Perth's development. The precinct gained ceremonial prominence in both political and civic discourse because of its name in honour of Queen Elizabeth II and the fact that it was built during a time of significant public infrastructure spending.

However, the idea of spatial inevitability—that Perth was destined to remain a city isolated from its river—was arguably the most significant story that Elizabeth Quay questioned. The project challenged this narrative and provided a fresh example of what public waterfront space should entail in an Australian city in the twenty-first century by rearranging land use, emphasizing pedestrian experience, and integrating cultural and recreational programs at the center of the reconstruction.

However, there were still some disputes surrounding the change. Concerns over land monetization and the dilution of truly public space were raised by critics who questioned the extent of private development integrated within the district. Even still, the project's importance may be found in this criticism: it compelled discussion regarding the rightful owner of the river and how it should be viewed.

Thus, Elizabeth Quay signifies a civic reawakening rather than just a shift in the cityscape. Rewriting the relationship between Perth and the Swan River, redefining the edge as a place of convergence rather than division, and reclaiming the public domain as the most potent narrative area in the city are all part of this endeavour. By doing this, it ushers in a new era of urban history in Perth, one where encounter, memory, and design come together at the water's edge.

6.1.1 Railways, Roads, and the Lost Edge

Perth's relationship with the Swan River was marked by distance rather than intimacy throughout a large portion of the twentieth century (Kennewell and Shaw 2008). The design of Perth's urban infrastructure gradually reduced both physical access and visual connection to the riverfront, despite the river's cultural, ecological, and symbolic importance to the city's identity (Cheshmehzangi 2020). What planners and urbanists now call a "lost edge", i.e., a threshold place devoid of its energy and civic meaning, was created as the center city, particularly the area south of St Georges Terrace, became operationally and perceptually separated from the water's edge.

The development of the railway and the predominance of motorized infrastructure marked the beginning of the disconnection. Early nineteenth-century Perth development was focused on the Swan River, which was the primary route for trade and transportation in the city. But over time, the building of railroad tracks and related infrastructure reoriented urban movements. Although these systems promoted economic development and regional communication, they also erected physical obstacles around the city. The division was further cemented in the 1930s when Riverside Drive was built as a significant east-west arterial highway. It served as a conduit for heavy traffic, physically obstructing pedestrian access to the river and transforming what would have been a civic promenade or linear park into a landscape dominated by cars.

This section of the river was dominated by parking lots, unused lawns, and pieces of recreational infrastructure by the post-war decades. Although it was inaccessible, the river was present in postcards and aerial photographs but was not a major part of daily life in the central business district. A planning rationale that prioritized vehicle throughput and divided land uses continued to marginalize even cultural and civic treasures, such the Supreme Court Gardens and the Barrack Street Jetty. The Swan River was reduced to a picturesque backdrop after serving as the spatial anchor of Noongar Country and early colonial life.

Decades of this situation were made worse by disjointed planning and a lack of investment in the public space along the shoreline. Perth was oddly detached from its river, in contrast to other Australian capitals. Due to physical obstacles and a conservative development philosophy that viewed the river's edge as residual space rather than a strategic opportunity, Perth fell behind other cities that had started to repurpose their riverfronts for civic use by the late 1990s and early 2000s, such as Brisbane and Melbourne.

With the introduction of Elizabeth Quay, a daring project spearheaded by the Western Australian Government and facilitated by the Metropolitan Redevelopment Authority (MRA), that started to alter in the early 2010s. The project's straightforward yet revolutionary tenet was to restore Perth's waterfront as a public and cultural resource. In order to physically bring the river back into the city grid and provide pedestrian access and visual continuity from important civic axes like William Street and The Esplanade, a portion of Riverside Drive was to be removed to construct an inlet.

Cutting into the landmass and reintroducing water was a calculated move that was more than just symbolic. It reframed Perth's waterfront as an extension of the city's civic heart rather than as unused land that needed to be manicured. The master concept called for a mixed-use district that would incorporate public art, commercial development, hospitality, walkable infrastructure, and event spaces. A broad promenade and a curved pedestrian bridge that crosses the new inlet would tie everything together. In addition to creating additional links between the Elizabeth Quay train and bus stations, ferry ports, and active transit options, the transportation system was redesigned to give precedence to pedestrians.

Concerns about commercial encroachment on public land, worries about increasing traffic, and the loss of Riverside Drive were some of the main issues surrounding Elizabeth Quay's development. But the project's political vision was unwavering: Perth could no longer afford to treat her river like a decorative feature. It had to become a living, breathing, and essential component of the city. The realization that a city's most valued boundaries are determined by their ability to support public life, cultural memory, and shared experience rather than just its infrastructure made Elizabeth Quay a landmark in urban planning. The goal of the project was to reestablish the city's most significant civic relationship—that between people and place, land and water—by reversing decades of estrangement, in addition to creating something new.

6.2 The Geometry of Connection: Design Thinking and Spatial Reunification

The idea behind Elizabeth Quay's reconstruction is straightforward but impactful: by reorganizing the spaces between, the city will be drawn closer to its river. In order to accomplish this, a fundamental rethinking of Perth's spatial geometry—how people move, halt, interact with water, and conduct public life along the urban edge—was necessary, in addition to the removal of roads and the construction of promenades. As a result, the precinct exhibits a thorough coordination of connection, design coherence, and public amenity, even if it is still being development.

The excavation of a new inlet from the Swan River onto the city's border is a distinctive feature of the Elizabeth Quay concept, which was created by ARM Architecture and landscape architects Taylor Cullity Lethlean (TCL). This water element is structural rather than decorative. By integrating the river into the constructed fabric and establishing a recognizable spatial axis between the downtown area and the waterfront, it restores a feeling of urban porosity. The river becomes a vital link—an arm of water that encircles civic, commercial, and cultural life—instead of a terminal.

This controversial but revolutionary architectural decision required the elimination of a significant portion of Riverside Drive. Once serving as a barrier between the city and the lake, the traffic corridor was redesigned to prioritize pedestrian and multimodal access while rerouting car flow. It has now been replaced by a winding public promenade that rims the inlet and is intended for both casual walking and heavy foot traffic during events. With views of the Swan River and the city skyline beyond, the pathway is interspersed with plazas, lookouts, and terraced seating.

The Elizabeth Quay pedestrian bridge may be the most recognizable intervention from a visual standpoint. It spans the inlet with a boldness that conveys civic confidence thanks to its dual-arched design, which is both structural and sculpture. More than just a piece of infrastructure, the bridge serves as a spatial anchor that unites the quay's northern and southern sides, frames important sightlines, and provides a viewpoint that allows one to appreciate the precinct's geometry. Additionally, the bridge encourages mobility loops, promoting walking, jogging, and cycling as regular urban activities. In contrast to the city's ancient colossal architecture, its design conveys a lightness that suggests a more fluid, modern concept of public space.

Elizabeth Quay's urban grain is purposefully delicate, especially in the way it combines formal civic rites with intimate moments of leisure. Because public areas are layered, they can be used for both planned and unplanned purposes. Without depending on strict borders, paving materials change in texture and pattern to denote various spatial uses, such as pathways, rest areas, and performance spaces. Stepped terraces, floating pontoons, and shade structures are some examples of water's edge treatments that are designed to encourage a tactile connection with the water. In addition to viewing the river, people are encouraged to sit next to it, touch it, and interact with it.

The curvature, permeability, and rhythm of Elizabeth Quay contrast with the hard-edged modernism of Perth's previous public architecture. To provide ample public circulation, buildings are set back. Retail, cafes, and cultural programming enliven their ground levels, creating clear boundaries between interior and outdoor spaces. With planting techniques that combine natural species with sculptural shape, the landscape is given equal consideration, resulting in shade, aesthetic appeal, and ecological usefulness.

The way the masterplan handles size is one of its more subtly effective features. The public realm is nevertheless human-scaled even though Elizabeth Quay is surrounded by towers and has significant development footprints, such as hotels, office buildings, and residences (Wood-Gush et al. 2015). The precinct doesn't feel overbuilt thanks to the openness of major spaces, the intimacy of seating areas, and the breadth of pathways. For the quay to feel like a place to live rather than just pass by or take pictures, the density and openness must be balanced.

Multimodal integration significantly improves connectivity. There are convenient pedestrian connections connecting transportation hubs to the shore, and Elizabeth Quay is immediately serviced by train, bus, and ferry infrastructure. There is obvious signage and wayfinding, and the landscape's visual cues aid in directing natural movement. As a result, the precinct feels open, accessible, and a part of the city's everyday rhythms—not a remote area, but a vital component of urban life.

Water itself is an active design element rather than only a background. The edges allow for fishing, observing, and contemplation; the entrance encourages kayak and boat activity. The central plaza's water features offer cooling and sensory stimulation, especially in Perth's sweltering summer months. These actions provide not only views of water but also actual interactions with it, reestablishing the city's connection to the natural world. All things considered, Elizabeth Quay's geometry of connection is both physical and symbolic. It connects form and function, intimacy and spectacle, city and river. The precinct uses spatial intelligence and strategic design interventions to reposition the river as a civic invitation rather than a border, creating a fluid interface that allows the city to see, touch, and rediscover itself.

6.3 Public Life on Display: Culture, Movement, and Placemaking in Practice

Elizabeth Quay was intended as a civic theater—a carefully planned and dynamic space for public life to unfold—rather than just an urban reconstruction. The real test of a precinct's vitality is how it functions over time—how it is used, who it attracts, what it hosts, and how it adjusts to both ordinary life and extraordinary occasions—even though its physical interventions are essential to its success. In this way, Elizabeth Quay has become one of Perth's liveliest public areas, providing a multi-layered experience of contact, culture, and movement.

Its carefully planned programming spanning time—day and night, weekday and weekend, summer and winter—is what gives it its vitality. Elizabeth Quay was planned and run to accommodate a steady flow of public life, in contrast to many waterfront developments that fall into a slumber outside of event times. The promenade draws visitors, families, joggers, and commuters during the day. The open seating spaces are used for casual meetings or coffee by office workers from neighbouring towers. Movement becomes experience thanks to dynamic water features, digital displays, and public artwork that provide reasons to stop, look around, or get involved.

After-dark vibrancy is also accommodated in the precinct, with late-night programs and lighting strategies being crucial. The quay feels lively and accessible far into the evening thanks to ground illumination incorporated into the walkways and surrounding the inlet edge, which also adds to the atmosphere and sense of safety. The precinct's infrastructure and visual openness are fully utilized during seasonal events like the Winter Lights Festival and New Year's Eve celebrations. Temporary artworks, soundscapes, and large-scale projections turn the waterfront into an immersive space that combines location and spectacle.

Elizabeth Quay hosts both planned and unplanned cultural events. A committed management team plans major events throughout the year, frequently in collaboration with state and local organizations, including markets, music performances, food and wine festivals, and cultural celebrations. These events draw crowds and help shape the identity of the quay as an urban destination. Smaller-scale, unofficial acts of occupation, such as couples having a picnic on the grounds, kids dancing in the water jets, or street vendors gathering crowds, are just as significant. The precinct feels genuinely public because of these spontaneous civic engagements that anchor it in daily life.

The incorporation of Indigenous placemaking and cultural expression is one of Elizabeth Quay's most unique features (Potter 2020). Cultural memory is incorporated into the site's modern form through artwork and landscape design components that are inspired by the stories of the Whadjuk Noongar people. For instance, dual-language allusions are included in the signage, and the inlet's design mimics the contours of historic landforms. Indigenous artists' public works of art, like "First Contact" by Laurel Nannup, explicitly address history of encounter, colonization, and persistent presence. These gestures, which promote a deeper engagement with place beyond superficial aesthetics, are essential to the precinct's spatial storytelling rather than being symbolic afterthoughts.

Additionally, commercial activation strategically supports public life. The promenade is flanked by a variety of eateries, cafes, galleries, and shops that provide places to stay longer and build multi-layered experiences. Notably, these commercial areas are thoughtfully interspersed with public areas to prevent the district from becoming overly exclusive or commodified. The steps that descend to the inlet, the grassy hillocks that provide lofty vistas, and the sculptured shade structures that provide relief on hot days are really some of the most popular parts of the quay that are open to the public.

Another essential component of the precinct's vibrancy is the activation of the water's edge. Elizabeth Quay encourages engagement, in contrast to riverfronts that perceive the water as inviolable. People hover near the water's edge to enjoy the breeze, see reflections, or feed the birds, while boats and ferries come and go from the inlet, and paddle boats and other small boats are scattered over the water. By serving as a circulation loop that promotes constant interaction with both land and water, the pedestrian bridge contributes to this layer.

As a result, Elizabeth Quay transforms from a location into a performance of city life. It demonstrates how a well-planned precinct can accommodate a variety of events for people of all ages, backgrounds, and schedules. Elizabeth Quay exemplifies the potential of placemaking as a living, participatory practice—where mobility, memory, and meaning intersect at the city's edge—by supporting both the extraordinary and the everyday.

6.4 Long-Term Urbanism: Legacy, Critique, and the Futures of Perth's Waterfront

This concluding section reflects on Elizabeth Quay's significance as a long-term urban catalyst. While drawing broader conclusions for planning waterfronts that integrate density, destination-making, and cultural inclusion in mid-sized global cities, it takes into account criticisms regarding commercialization, authenticity, and gentrification. Today, Elizabeth Quay is not only a finished project but also a work-in-progress, a crucial step in the long-term urban development of Perth. As a symbol, it stands for a shift in civic priorities: a deliberate focus on urban density, public space, and cultural presence at the most symbolic edge of the city. As a precinct, it provides a concrete re-evaluation of how placemaking, waterfront design, and infrastructure might be combined to bring the river back into the mainstream of urban life. But like all well-known urban initiatives, Elizabeth Quay's legacy is complex, containing both successes and unresolved conflicts.

The worry about the commodification and commercialization of public land is one of the most prevalent criticisms. Large-scale private investment, branded hotels, and opulent developments have raised concerns about whether the public realm has actually been maintained. Even though the concept includes open public access and ample pedestrian spaces, proponents of inclusive urbanism are nevertheless concerned about the impression of elite capture, particularly through waterfront high-rises. In this regard, Elizabeth Quay is a sobering example of the need to carefully preserve throughout time the harmony between private growth and public philanthropy.

Authenticity and local identity are other points of criticism. Despite the precinct's modern and globally recognizable architectural language, some critics contend that it runs the risk of seeming out of place since it is more in line with waterfront aesthetics around the world than it is with Perth's distinct cultural and environmental heritage. This has been partially resolved by the incorporation of Indigenous stories and regional art, but it is still difficult to guarantee that long-term programming, spatial use, and symbolic representation continue to represent the city's changing and varied identity instead of falling back on a developer-led or tourism-centric vision.

Elizabeth Quay has already sparked significant changes in Perth's views on civic space and urban development in spite of these conflicts. It has sparked a resurgence of interest in mixed-use development, walkability, and the imaginative possibilities of water-edge landscapes. Perth's urban approach has also matured as a result, establishing the city as one that can take risks, think long-term, and invest in areas that are truly experiential rather than just transactional. Looking ahead, Elizabeth Quay's success will depend on how adaptable and inclusive it stays as it develops. Its public areas must continue to be flexible enough to accommodate new populations, climates, and uses. Its cultural offerings need to continue to represent the diversity of Perth's communities (Jones and Newsome 2015). Additionally, its governance needs to guarantee that placemaking (Cheshmehzangi 2016) is maintained as a public interest rather than only a design element.

Elizabeth Quay is a useful paradigm for other mid-sized towns struggling with post-industrial waterfronts, not as a model to follow but as a precedent to consider. It teaches that only with persistent vision, interdisciplinary cooperation, and a dedication to the commons can there be a reconnection—between city and river, past and future, design and everyday life. Elizabeth Quay is thus the start of a more urban, involved, and interconnected future for Perth rather than the conclusion of a process.

References

Andrews S, Stanley F, Eades S (2024) First knowledges health: spirit, country and culture. Thames & Hudson Australia

Cheshmehzangi A (2016) Temporary and temporality: public realm regeneration through temporary events. J Urban Regene Renew 10(1):58–72

Cheshmehzangi A (2020) Reflecting on urban identities. In: Identity of cities and city of identities, Springer, Singapore, pp 133–178

Cooper C (2015) Best in urban renewal. Prop Aust 29(2). Available from: https://doi.org/10.3316/informit.755813343096436

Critchley A (2022) The politics of eternity: public history and monuments to memory in Kings Park, 1902–1934. Available from: https://researchonline.nd.edu.au/theses/349/

Jones C, Newsome D (2015) Perth (Australia) as one of the world's most liveable cities: a perspective on society, sustainability and environment. Int J Tour Cities 1(1):18–35

Kennewell C, Shaw BJ (2008) Perth, Western Australia. Cities 25(4):243–255

Potter E (2020) Contesting imaginaries in the Australian city: urban planning, public storytelling and the implications for climate change. Urban Stud 57(7):1536–1552

Wood-Gush B, Ciemitis P, Egan J (2015) Placing the citizen in the centre: the rise of great places in Western Australia. Plan News 41(4):10–11

Part II
CBD Expansion and High-Density Urban Growth

Major urban densification and city-building projects that reshape skylines and improve connectivity:

- **Parramatta Square, Sydney**—A key expansion of Sydney's second CBD, integrating commercial and civic spaces.
- **Collins Square, Melbourne**—One of Australia's largest mixed-use business precincts.
- **Maroochydore City Centre, Queensland**—A smart-city-integrated new CBD.
- **Brisbane Live Precinct, Brisbane**—A major entertainment and commercial hub proposal for redefining the city skyline.
- **Civic Quarter, Canberra**—A contemporary office and public precinct reshaping the capital's urban core.

Chapter 7
Building a New Centre: Parramatta Square and the Making of Sydney's Second CBD

7.1 Divided by Design: The Historical Disconnection of Perth and the River

Parramatta Square is a striking spatial statement in the restructuring of Greater Sydney, not just a reconstruction. One of Australia's biggest urban regeneration initiatives, Parramatta Square is located in the center of Sydney's designated "Central City" and symbolizes a radical reimagining of high-density civic and commercial activity outside of the conventional CBD (Rauscher et al. 2017). This chapter examines how integrated design, vertical intensity, public realm extension, and a compelling civic narrative were used to reinvent a previously fragmented city center, opening up new possibilities for decentralized urban growth in Australian cities.

Decades of population growth and evolving urban logic have led to Parramatta's rise as Greater Sydney's geographic and strategic center. Parramatta, which was once thought of as a secondary administrative hub that was both vital and ancillary, is today acknowledged as a metropolitan core in and of itself. Together, its institutional presence, transportation hub, cultural diversity, and population expansion have transformed it from a satellite city into Sydney's "Central River City." In order to alleviate the strain on the Eastern Harbour CBD, promote economic equity throughout the basin, and bring employment, housing, and services closer to the residents of Western Sydney, the Greater Sydney Commission's "Metropolis of Three Cities" vision enshrined this role as both a necessary urban adjustment and a policy goal.

In this sense, Parramatta Square becomes a tangible representation of this change in urban order rather than just an architectural precinct. It functions as a rebalanced city's commercial and civic center. The project's magnitude and purpose are what make it unique. Parramatta Square is the result of a long-term, staged vision—consciously bringing together government, private developers, and community-facing organizations around a similar urban outcome, as opposed to the ad hoc densification typical of many second CBDs. With its completion, Parramatta reaches

A. Cheshmehzangi, *Australia's Urbanity*, Urban Sustainability,
https://doi.org/10.1007/978-981-95-3881-2_7

a stage of urban maturity where coherence, public value, and a significant sense of place are used to gauge progress rather than just increase.

7.1.1 From Administrative Centre to Metropolitan Core

The transformation of Parramatta from a colonial administrative outpost to a vibrant metropolitan center is indicative of larger changes in Sydney's approach to urban development. Parramatta has historically functioned as a secondary center, housing suburban residential areas and administrative activities. But its course underwent a significant shift at the start of the twenty-first century.

Through the Greater Sydney Commission, the New South Wales Government unveiled the "Metropolis of Three Cities" idea in recognition of the need to handle Sydney's expanding population and economic activity. With the help of this strategic plan, Greater Sydney was to be divided into three cities: the Western Parkland City, the Central River City (Parramatta), and the Eastern Harbour City. According to this plan, Parramatta was supposed to be the Central River City, a center that supports the growing Western Sydney area while counteracting the traditional CBD's draw.

The creation of Parramatta Square, a $3.2 billion urban revitalization project covering a three-hectare plot of land in the center of Parramatta (Langley 2013), was essential to this idea. Under the direction of Johnson Pilton Walker (JPW), the project's master plan sought to establish a unified precinct that combines public, commercial, and civic areas. 290,000 m^2 of commercial space and 6,000 m^2 of public space are provided by the development's six structures, which include Australia's largest commercial office complex and a renovated Town Hall.

Parramatta Square's advantageous location next to important transportation hubs, such as the Parramatta train station and the upcoming Parramatta Light Rail, improves accessibility and solidifies its position as a key hub in Sydney's transportation system. The precinct's ability to house a projected 23,500 workers is supported by this connectedness, making it a strong contender to replace the conventional CBD. Moreover, the integration of cultural and educational institutions, such as the University of Western Sydney's vertical campus and the PHIVE civic hub, enriches the precinct's urban fabric. A thriving public environment is fostered by PHIVE, a communal, cultural, and civic center that includes a library, maker spaces, tech labs, and community meeting rooms.

A purpose-led approach to urban planning is demonstrated by the renovation of Parramatta Square, where accessible public areas and civic facilities coexist with high-density building. This concept offers a plan for equitable and sustainable urban growth, challenging the conventional monofunctional CBD paradigm. In summary, the growth of Parramatta Square represents a deliberate change in Sydney's urban design that embraces polycentricity and decentralization. The possibilities of high-density urbanization in Australia's metropolitan landscape are redefined by Parramatta Square, which combines commercial energy with civic involvement and public accessibility.

7.2 Density with Purpose: High-Rise Form, Public Space, and Urban Logic

The architectural, infrastructure, and spatial design techniques incorporated into Parramatta Square are examined in this section. In order to provide a new model for human-scaled high-density urbanism in Australia, it will examine how vertical growth has been combined with a sizable public realm, integrated transport connections, activated street frontages, and smooth transitions between skyscrapers, plazas, and civic institutions.

Parramatta Square serves as an example of how vertical density may enhance rather than detract from public life when it is based on a unified urban logic and civic vision. Parramatta Square adopts an urban paradigm that strikes a balance between ambitious height and openness, scale and permeability, and intensity and inclusivity, in contrast to much of Sydney's older high-rise building, which frequently placed an emphasis on economic output over experiential quality. This equilibrium is not a result of chance; rather, it is ingrained in the planning and design logic that shape the precinct.

A 7,000-square-meter public plaza, the largest of its sort outside of Sydney's central business district in New South Wales, serves as the focal point of the complex. This large square serves as the development's centerpiece, allowing for easy pedestrian traffic and social interaction in addition to serving as an open area. Because of careful massing, set-backs, and progressive edge treatments, it serves as both foreground and backdrop, surrounded by towers that reach over 50 stories without ever being visually overpowered. As a result, the vertical urbanism still reads from the ground at a human size.

Major commercial buildings (6, 8, and 10 Parramatta Square) and civic and institutional architecture, including the renovated Parramatta Town Hall and PHIVE, a remarkable community and library hub, are also part of the tower cluster. All of these elements work together to create a precisely calibrated spatial choreography. For instance, Johnson Pilton Walker's design for 6 and 8 Parramatta Square has podiums with lively frontages, shops, and covered arcades that promote interaction with the plaza rather than architectural seclusion. The towers' mass is raised on slender columns, which produces a visual permeability that lets sightlines, movement, and light into the center of the precinct.

The density of Parramatta Square is unique because of its porosity, which allows traffic to flow through, across, and between buildings. The complex is accessible from all directions because to its numerous entry points, pedestrian lanes, and broad promenades. What could otherwise be a crowded business district can serve as a municipal common because to its permeability. The purpose of Parramatta Square is to facilitate continuous activity, in contrast to corporate plazas that close after business hours. To foster what urban thinker Jan Gehl referred to as "life between buildings" (Gehl 2011), ground-floor eateries, adaptable lounging places, outdoor art exhibits, and landscaped pauses all collaborate.

One important enabling factor is transportation infrastructure. Directly next to the square is Parramatta Station, a multimodal junction with integrated bus service, regional train lines, plans for metro development, and the future Parramatta Light Rail. These transportation connections are integrated into the precinct's urban fabric rather than being tucked away, which reduces obstacles and increases foot traffic. There is a strong sense of spatial logic created by the smooth transition between the plaza and transit: this is a location made for mobility, arrival, and exchange. More significantly, this transit-oriented approach strengthens the precinct's reputation for urban sustainability and lessens its reliance on cars.

The square's surface design exhibits consideration for its social and environmental purposes. While tree canopies and water features provide comfort during Parramatta's scorching summers, high-albedo materials help to cool the city. Seating is positioned throughout the area, not only at the perimeter of the plaza, to promote loitering and conversation. Stormwater management has been achieved through the use of water-sensitive urban design techniques, and nighttime vibrancy is maintained by energy-efficient lighting systems. Together, these micro-interventions create the infrastructure of liveability, even though they are frequently inconspicuous when in use.

Integrating civic institutions into the development has been one of the most crucial strategic decisions. With maker spaces, meeting spaces, art exhibits, and public services, PHIVE (5 Parramatta Square) is more than just a library. Its design, created by the French firm Manuelle Gautrand Architecture in partnership with Lacoste + Stevenson, provides a translucent and sculpture-like contrast to the surrounding steel and glass structures. A strong commitment to place-led development is demonstrated by this civic insertion in the square's very center: density as a venue for community life rather than as a financial commodity.

The urban logic of Parramatta Square is also evident in how edges have been treated (Barns and Pollio 2018). Instead of having distinct borders, the area carefully blends into the surrounding streetscapes. There are clear and welcoming connections to Church Street, the riverbank, and nearby cultural and educational institutions. This multi-tiered strategy aids in integrating the district with Parramatta's larger urban fabric, preventing it from being perceived as an isolated area but rather as an extension of the city's lifeblood.

In conclusion, the high-density paradigm is reframed by Parramatta Square. It repositions verticality inside a civic-centered, human-scaled spatial framework rather than rejecting it. It illustrates how infrastructure may promote openness, how physical form can support public life, and how design can unite density with dignity. Parramatta Square is a prime example of purposeful density, where architecture, urban processes, and daily living are all part of a single urban ethic rather than distinct agendas, as cities around Australia continue to grow higher and outward.

7.3 The Everyday City: Activation, Culture, and the Rhythms of Life

If Parramatta Square's architecture serves as its framework (Parramatta City Council 2014), the daily activities that take place there serve as its artwork. Parramatta Square is more than just a striking arrangement of towers and plazas; it has been designed as a civic arena where the fusion of government, culture, leisure, and employment is not only controlled but also honoured. Its physical shape is simply one aspect of its success as an urban design element; another is its capacity to maintain an inclusive, continuous, and context-and time-adaptive rhythm of activity.

The intentional layering of functions is at the core of this vibrancy. In a closely knit urban grid, Parramatta Square houses government buildings, business headquarters, retail establishments, cultural institutions, and educational facilities. Over the course of a day or week, the square is animated by the various flows created by these strata, including those of employees, students, families, tourists, artists, and officials. During lunch, office workers overflow into the plaza. Western Sydney University students frequently use the common areas in between classes. Families from the area come to public events or weekend markets. With their diverse schedules and space requirements, these user groups provide a textured rhythm that defies monoculture.

The cafes and eateries that line the border of the plaza are essential to this cadence. Their extended operation hours, outdoor seating, and street-facing location contribute to extending activity beyond the typical 9–5 weekday. Crucially, the precinct has escaped turning into a sterile business district. Alongside more permanent, carefully chosen products, the area frequently includes pop-up shops, food trucks, and small-scale vendors that provide spontaneity and affordability. These retail characteristics add to the precinct's "slow vibrancy," which can be described as a consistent, hospitable, and human-scale vitality.

Programming has been crucial to fostering this vitality. Parramatta Square serves as a venue for public life in addition to being a site of transit. Frequent activities guarantee that the space is not just used but also experienced, from outdoor exhibitions and civic commemorations to live music and cultural festivals. In order to reflect the rich cultural diversity of Western Sydney, special attention has been given to multicultural and Indigenous representation in programming. Diwali celebrations, Lunar New Year celebrations, and NAIDOC Week installations are not extracurricular activities; rather, they are a part of the precinct's yearly routine, highlighting its significance to the communities it serves.

In my experience, digital public art is innovative (Cheshmehzangi 2023). Large-scale LED screens, projection mapping on building façades, and interactive installations offer contemporary interpretations of public storytelling. The distinction between art and infrastructure is blurred by these media-based interventions, which offer dynamic content that adapts to social, political, and cultural situations throughout the year. These digital interventions enhance public interaction by enabling Parramatta Square to modify its story and aesthetic in real time, in contrast to traditional public art that stays static.

The precinct's strategy for activation at night is equally excellent. A feeling of coziness and inclusivity after dark is influenced by lighting design, safety features, and the allocation of evening-oriented usage. A vibrant and family-friendly nighttime economy is supported by late-opening restaurants, civic night activities, and digital light installations. This is a significant change from typical central business districts, which frequently turn dreary or are overrun by nightclubs after hours. Another indication of Parramatta Square's place-led attitude is the evening use, which is less about consumption and more about ongoing civic engagement.

Importantly, public government is not separated from the public sphere because to the co-location of civic institutions inside the precinct, including the City of Parramatta's offices, the public library within PHIVE, and nearby law courts. On their way to cast their ballots, go to council meetings, or get services, citizens pass through the square. This is creating cities with people, not just for them. Accessibility and civic transparency transcend bureaucratic ideals and become architectural and geographical norms.

Parramatta Square's ability to conflate conventional urban dichotomies is one of its greatest advantages. Work and leisure are not sharply separated; neither are education and culture, or government and commerce. The precinct serves as an example of how urban space may accommodate overlapping purposes without creating conflict by anticipating diversity, promoting coexistence, and building for flexibility. Routine and ritual, daily life and ceremonial spectacle, casual get-togethers and institutional functions are all accommodated in this precinct.

Parramatta Square serves as an example of how densely populated metropolitan areas may maintain its human qualities in this way. It is a new type of CBD, not just a new one. One that evolves, invites, listens, and adjusts. Its vibrancy is fostered by ongoing social investment, spatial care, and cultural responsiveness rather than being imposed by branding. In every way, it is a typical city that is pulsating with its citizens.

7.4 Designing for a New Urban Identity: Lessons from the Heart of Western Sydney

Parramatta Square is the architectural and metaphorical focal point of a new metropolitan identity for Western Sydney, and it is more than just the collection of buildings and plazas. Urban planners, designers, and community leaders have done more than just renovate a piece of land to create Parramatta Square. They have created a civic narrative that emphasizes equity, decentralization, cultural pride, and the pressing need for inclusive and expanded urban growth. However, its emergence is surrounded by paradoxes, difficulties, and disputed meanings, much like every revolutionary urban initiative.

Parramatta Square is an example of what a polycentric metropolis can look like as Greater Sydney's gravitational center moves westward. Unlike inward-facing commercial parks or haphazard suburban expansion, the Square makes a strong case for high-density urbanization that is cohesive, connected by transit, and civically oriented. It demonstrates how public accessibility may be maintained while government buildings, business towers, cultural institutions, and community facilities coexist in close vertical and horizontal proximity. It challenges upcoming urban centers in Australia to consider the notion that deliberate density design can strengthen rather than weaken a city's social cohesion.

However, there are conflicts associated with this achievement. The fast rise of land values, rental costs, and redevelopment pressures that have accompanied Parramatta's change have raised long-term worries about affordability and displacement. Although the Square is open to the public, the surrounding area is becoming more and more characterized by private investment and speculative real estate reasoning. A fundamental question is brought up by this dynamic: will Parramatta Square's civic spirit survive if the city as a whole grows more economically exclusive? It will take constant governance, policy attention, and design vigilance to maintain real public benefit—rather than symbolic openness—as more developments congregate around the Square.

Another point of contention is cultural representation. One of Australia's most linguistically and ethnically diverse cities is Parramatta (Nectoux 2021). It is not always easy to translate that diversity into everyday life and built form, though. Deeper cultural embedding is still developing, even though the Square's events programming has made progress in presenting Indigenous stories, multicultural festivals, and public art. Beyond celebration, true inclusion must address how many communities access decision-making, shape space, and have their history represented in programming, materials, and architectural languages.

However, the fact that Parramatta Square shows that urban identity may be created from scratch rather than passed down is among its most significant accomplishments. The colonial infrastructure, elite financial institutions, and planning paradigms that favour east-facing expansion all weigh heavily on Sydney's original central business district. On the other hand, Parramatta Square is a reinterpretation. It presents a fresh perspective on Australian urbanity, one based on openness to the public, daily energy, and generous use of space. Its identity is purposefully modern while also being sensitive to the environment, local heritage, and the cultural life of its residents.

The Square also shows that placemaking and infrastructure are intertwined from a design standpoint. Transit, water systems, digital connection, and energy methods are not concealed beneath the urban surface; rather, they are manifested in the experience of moving through space, ingrained in the public realm, and articulated in the logic of the streets. A municipal square can serve as both infrastructure and identity (Cheshmehzangi 2020), as demonstrated by the arrangement of public services, transportation options, cultural landmarks, and adaptable gathering places.

To sum up, Parramatta Square represents a sophisticated shift in Australian urban planning. It acknowledges that decentralisation is not about merely building new towers in new locations, but about creating places of shared meaning, economic opportunity, and cultural richness. It also shows that public space is still a potent instrument for forming a common identity, particularly when it is intended to serve as a canvas for daily life rather than a stage for special occasions. A new urban narrative is being written in the center of Western Sydney, one in which the city center is created rather than inherited, architecture fosters community, and density promotes inclusiveness.

References

Barns S, Pollio A (2018) Parramatta smart city and the quest to build Australia's next great city. In: Inside smart cities. Routledge, New York, pp 197–210

Cheshmehzangi A (2020) Identity of cities and city of identities: examples. In Identity of cities and city of identities. Springer, Singapore, pp 179–244

Cheshmehzangi A (2023) City life experiences: lessons learnt for future urban regeneration strategies. In: Mapping urban regeneration: city life experiences in Yunnan, China. Springer Nature Singapore, Singapore, pp 239–246

Gehl J (2011) Life between buildings: using public space. The Danish Architectural Press, Copenhagen

Langley J (2013) New funding options for urban renewal. In: 49th ISOCARP Congress, SKM, Australia

Nectoux SA (2021) Planning cultural infrastructure in the multicultural "city within the city": understanding attitudes and perceptions of cultural spaces of young adults from Indian and Chinese backgrounds in the City of Parramatta. Master of Research, Thesis for Western Sydney University. Available from: https://researchers-admin.westernsydney.edu.au/ws/portalfiles/portal/94924752/uws_67005.pdf

Parramatta City Council (2014) Parramatta CBD planning framework: economic analysis. By SGS Economics & Planning. Available from: https://www.ipcn.nsw.gov.au/sites/default/files/pac/projects/2018/09/request-for-gateway-review-of-land-at-55-aird-street-parramatta/additional-information-from-council/draft-report--sgs--parramatta-cbd-planning-framework-study--aug-2014_1.pdf

Rauscher RC, Momtaz S, Rauscher RC, Momtaz S (2017) Planning in Greater Sydney–Parramatta City case study. In: Cities in Global Transition: Creating Sustainable Communities in Australia, pp 179–198

Chapter 8
Vertical Synergies: Collins Square and the Fabric of Future CBDs

8.1 A City Within a City: The Rise of Collins Square

Melbourne's ambitious plans for high-density development and urban renewal are exemplified by Collins Square. This large mixed-use business zone, which is located in Docklands' Batman's Hill precinct, has turned a neglected area into a bustling center of retail and commercial activity. One of Australia's biggest commercial complexes, Collins Square was created by Walker Corporation (Whittle 2016) and includes six office towers, the historic Goods Shed South that has been renovated, and more than 10,000 m^2 of retail space.

Collins Square's creation fits into Melbourne's larger goal to rehabilitate the Docklands neighbourhood, which was formerly defined by abandoned industrial buildings and a dearth of well-thought-out urban planning (Magdy Shafiq Mohamed 2022). The foundation for this change was established by the Docklands' Conceptual Planning and Design Framework, which was published in 1996 and placed a strong emphasis on the integration of public, residential, and commercial areas. With the goal of bridging the divide between Melbourne's old central business district and the growing western precinct, Collins Square developed as a landmark project within this framework.

Collins Square's architecture combines contemporary style with historic preservation. Reputable companies including Bates Smart, Buchan, Hassell, and Woods Bagot developed the precinct's towers, which have modern amenities and streamlined facades. In addition to preserving a portion of Melbourne's industrial past, the development's incorporation of the historic Goods Shed South gives the precinct a distinctive personality. This blend of the ancient and the new is a perfect example of Melbourne's approach to urban development, where modernity and history preservation coexist together.

Collins Square is important to Melbourne's economy in addition to its architectural value (Davison 1998). Numerous commercial tenants occupy the precinct, providing a vibrant business environment and stimulating regional economic growth.

A. Cheshmehzangi, *Australia's Urbanity*, Urban Sustainability,
https://doi.org/10.1007/978-981-95-3881-2_8

Being close to important transportation hubs like Southern Cross Station guarantees smooth connectivity, which makes it a desirable location for both employees and enterprises. Collins Square is essentially the successful completion of a mixed-use business district that not only satisfies the needs of contemporary trade but also adds to Melbourne's urban fabric. Its creation serves as a reminder of how crucial innovative architecture, strategic planning, and the use of historical features are to creating thriving, sustainable urban landscapes.

8.1.1 Strategic Foundations and Urban Ambitions

A combination of strategic planning projects and urban aspirations to revitalize Melbourne's Docklands district served as the foundation for Collins Square's growth. A driving force behind the transformation of the abandoned shoreline into a vibrant extension of the central business district was the Docklands' Conceptual Planning and Design Framework of 1996. This framework placed a strong emphasis on public space creation to encourage community involvement, mixed-use development, and interaction with existing urban areas.

These strategic goals were in line with Walker Corporation's purchase and subsequent development of the Collins Square property. In order to promote economic activity and urban vibrancy, the business aimed to establish a "city within a city," combining retail establishments, public amenities, and commercial office spaces. This strategy aligned with the larger objectives of the Docklands redevelopment, which aimed to relieve strain on the historic CBD and decentralize Melbourne's business activity.

Collins Square was made possible in part by the Victorian government's policy backing. Large-scale projects were made possible by the state's dedication to infrastructure development and urban redevelopment. The development of Collins Square relied heavily on the creation of high-quality, easily accessible public areas, which were highlighted by initiatives like the Docklands Public Realm Plan.

Furthermore, Melbourne's urban aspirations are reflected in Collins Square's incorporation of sustainable design concepts. With features like tri-generation systems, rainwater harvesting, and energy-efficient fixtures, the precinct has a high concentration of ecologically friendly buildings. These features improve the precinct's charm and liveability while also lessening its environmental impact. In summary, the urban goals and strategic underpinnings that shaped Collins Square's growth serve as an excellent example of an all-encompassing strategy for urban renewal. Melbourne has effectively turned a once-neglected area into a bustling, integrated business sector that makes a substantial contribution to the urban landscape of the city by coordinating legislative frameworks, private investment, and sustainable design.

8.2 Density Meets Design: Architecture, Integration, and Urban Legibility

An excellent example of how high-density, vertically scaled urban form can be coordinated into a unified, easily navigable, and human-centered precinct is Collins Square. Collins Square is an example of what may be achieved when architectural design, public realm planning, and circulation logic are closely interwoven from the beginning, as opposed to the fragmented development that frequently defines huge urban developments. It demonstrates that purposefully designed density need not be intimidating; it may be readable, understandable, and even graceful.

The architectural choreography of the five towers that make up Collins Square's skyline is key to this accomplishment. Although the form and articulation of each tower vary significantly, they are all connected by a common architectural language that strikes a balance between aesthetic restraint and commercial seriousness, a uniform material palette, and a cohesive massing. The towers define a sequence of internal volumes and physical edges that orient the visitor and direct movement, rather than visibly competing with one another. In order to create a comfortable interaction between the monumental and the daily, podiums and street-level treatments attenuate the vertical scale at the pedestrian level.

Importantly, a network of interconnecting retail and circulation concourses ties the towers together spatially; they are not separate monoliths. The precinct's "urban interior," a vast, weatherproof public space that guarantees permeability and navigability across the property, is made up of many internal passageways. Natural lighting, a range of ceiling heights, material transitions, and visual cues that aid in orienting and guiding visitors make the experience of moving across these concourses more than just transactional. What could have easily turned into a closed business space has been transformed into an open and dynamic space that is equally beneficial for visiting professionals, casual diners, and daily commuters.

Collins Square's legibility is further strengthened by using a fine-grained spatial sequencing technique. There are no sudden changes in the precinct's transitions between the public, semi-public, and private domains. Openness and control are respected by the way office tower entrances, retail establishments, and hospitality venues are layered. Whether from a café into an atrium or from the plaza into a lobby, sightlines across the precinct provide a feeling of scale and spatial coherence that makes the tall buildings around them more approachable. By allowing for a more dynamic interaction between people and the built form, this type of layering makes sure that activity is not limited to the outside.

Another essential element of the design logic is integration with the transportation infrastructure. Collins Square is positioned as a truly transit-oriented development due to its proximity to Southern Cross Station, one of Melbourne's major transportation hubs. High foot traffic is possible without any confusion or congestion thanks to the station's pedestrian flow being perfectly incorporated into the precinct's concourse layout. The district functions at both the city and street levels thanks to connections to tram lines, bike lanes, and subterranean parking that are all planned to reduce potential sites of contention.

The precinct's character is further enhanced by the incorporation of the ancient Goods Shed South into the new architectural ensemble. Instead of being treated as a museum exhibit, the building has been purposefully transformed into dining and business spaces, connecting the site's industrial history with its modern use. By providing a recognizable anchor amidst the glass and steel, this not only strengthens the precinct's urban legibility but also adds to its architectural diversity.

The precinct's open space has been thoughtfully designed to strike a balance between exposure and containment. To provide light, vegetation, and breaks, small plazas, forecourts, and breakout spaces are placed in between buildings. Despite their modest size, these areas are useful for creating breathing room in the crowded urban fabric. These areas are made more visually appealing and serve a more social purpose by the thoughtful placement of seating, shade, vegetation, and public art.

Collins Square stands out for its capacity to provide urban clarity in the midst of extreme complexity. It fosters a walkable, cohesive, and publically welcoming urban environment while managing the high programmatic needs of commercial development, such as efficiency, security, and flexibility. This is density with a design goal rather than density for its own sake. It proves that placemaking, human scale, and civic giving can all be found in corporate real estate settings. Collins Square provides an example of how to use spatial intelligence to densify metropolitan areas as Australia struggles with this issue. It demonstrates that scale and sensitivity may coexist, that verticality can be made readable, and that the experience of city life need not be subordinated to the economic imperative. Rather, it can improve it if it is well constructed.

8.3 Life in Motion: Everyday Urbanity and Public Experience

Collins Square is not a monolithic business enclave despite its size and commercial intensity; rather, it is a precinct that is energized by daily living, complex encounters, and unplanned urban engagement moments (Zakariya et al. 2016). Collins Square's meticulous and ongoing attention to urban animation, in addition to its architectural design, sets it apart from other high-density office spaces. The area is alive with activity, interaction, and exchange, from morning coffee rituals to after-work get-togethers.

The precinct's public interface is defined by a well-planned network of ground-level and mezzanine retail establishments. To accommodate the rhythms of metropolitan working life, cafés, restaurants, informal dining establishments, wellness centers, convenience services, and boutique retail stores have been carefully chosen and placed. These facilities are essential to Collins Square's overall experience and are by no means an afterthought to business leasing. By transforming hallways and lobbies into destinations rather than just transit areas, they facilitate the day's ebb and flow of activity.

From upscale business lunches to quick street cuisine, the food and hospitality options have also been carefully chosen to appeal to a wide range of tastes. The fast-paced schedules of weekday employees and the more relaxed interactions of tourists, visitors, and locals are both accommodated by this tiered offering. Collins Square promotes an inclusive kind of activation that keeps the precinct from seeming exclusive or monofunctional by providing several "speeds" of urban life in one area.

The precinct's ability to accommodate informal use is what makes it so effective. Collins Square offers areas that seem free to interpretation even though its commercial functions are well-defined. For example, the internal concourses are sufficiently broad and diverse to accommodate everything from spontaneous social gatherings to minor cultural activations to discreet meetings and mobile working. These interior spaces are transformed into social landscapes by the incorporation of vegetation, high ceilings, ample seating, and acoustic sensitivity.

A dedication to spatial generosity lies at the core of the precinct's public experience; this is especially crucial in a precinct that is characterized by verticality. Entry plazas, courtyards, and outdoor areas are all purposefully permeable and seamlessly integrated with the interior proportions. These spaces are quite practical and human-scaled, but they are neither opulent nor ceremonial. They offer chances for passing through, brief phone conversations, lunch breaks in the sun, and casual conversations. Their success stems from their ordinariness; rather than being overly contrived, they offer a peaceful framework that allows daily urban life to unfold.

Because of its proximity to Southern Cross Station, the precinct also enjoys consistent and steady foot traffic. Collins Square is never isolated because to this transport-linked urbanism. In addition to tenants and employees, thousands of individuals who commute between the Docklands and the central business district occupy its public areas, bringing in a constant flow of casual visitors. The district has an urban life outside of regular office hours because to its dynamic circulation, which also supports its retail economy.

In Collins Square, placemaking also has a cultural and digital aspect. Throughout the year, pop-up exhibits, seasonal programs, and light projections offer additional layers of sensory stimulation. Through community activations during significant city events and art exhibits in interior corridors, the area transforms from a workplace into an integral element of Melbourne's western edge's cultural landscape. By reminding visitors that the most successful areas in the city are those that permit surprise, play, and pause, these activations lessen the place's commercial feel.

Crucially, Collins Square's architecture encourages these encounters without being dictated. The architecture does not prescribe use; rather, it outlines options. In addition to facilitating efficient mobility, wide circulation channels permit loitering and gathering. Lighting, physical warmth, and visual links to the exterior enliven the lobby and atrium spaces, beckoning visitors to stay. The architectural character is able to communicate without overpowering the spatial experience because even the building signage and tenant branding are kept to a minimum.

Socially, Collins Square operates as a mixing zone. Despite being constructed mostly for office tenants, it draws commuters using the station, city workers taking lunch breaks, students from surrounding universities, and tourists from the larger Docklands area. By naturally sharing space, these overlapping groups produce the bustle of urban life that many corporate precincts find difficult to produce. This organic combination is the result of deliberate planning that places a high value on porosity, connectedness, and diversity in both programming and design.

Finally, Collins Square's greatest accomplishment lies not in its size or height but rather in its capacity to sustain moving life. People move, meet, live, and return there—not only to work, but to be a part of something more than their daily responsibilities. By doing this, it reimagines what a business district may be: a live, breathing part of the city rather than a closed-off collection of skyscrapers.

8.4 Lessons in the Vertical Commons: Rethinking Mixed-Use Precincts

A striking illustration of how high-density commercial development may transcend corporate containment and meaningfully interact with the larger city is provided by Collins Square. Its architecture, integration, and day-to-day operations demonstrate how private-led developments may create an urban feel that transcends their boundaries when they are directed by sound spatial reasoning. But in the context of future Australian city-making, Collins Square also encourages critical thought about the obligations, dangers, and opportunities of vertical precincts.

Collins Square's ability to provide a cohesive and manageable urban experience in spite of its size and programming density is among its most noteworthy achievements. The precinct is successful in coordinating a number of skyscrapers, interior concourses, and outdoor areas into a recognizable setting that facilitates social interaction, daily rhythms, and pedestrian circulation. In a period when many cities struggle to balance commercial density with public life, Collins Square demonstrates that spatial generosity, even in small gestures—like atrium connections, outdoor plazas, or seating nooks—can soften vertical intensity and humanise the high-rise landscape.

Its placemaking concept, which consists of tiered facilities, carefully chosen ground-plane retail, and regular activation, teaches us a valuable lesson: vibrancy needs to be both programmed and allowed. Collins Square depends on more than just natural urban activity. Rather, through varied shop spaces, adaptable public spaces, and small-scale interventions like lighting and seasonal activities, it intentionally encourages and promotes life. A key factor in creating effective high-density areas is that the precinct seems used rather than merely constructed.

The success of Collins Square, however, also calls into question the boundaries of publicity in privately held areas. Even though the forecourts and concourses are open and often used, private management nonetheless controls them. This presents possible limitations on the types of social gatherings, protests, and expression that characterize authentic civic space. It is important to consider how concepts of accessibility, democratic use, and inclusivity are maintained throughout time as more Australian towns adopt vertical precincts created by commercial organizations.

Care must be taken while striking a balance between civic openness and economic control. A hybrid model—one that strives for public ideals in experience but is not entirely public in ownership—may be suggested by Collins Square's generous design and public invitation. However, this is still dependent on developer vision and management ethos, which might differ significantly amongst projects, in the absence of defined legislative frameworks or performance goals for the public good.

Making sure that vertical mixed-use precincts adjust to changing urban needs presents another difficulty. Traditional office use patterns were upset by the COVID-19 pandemic, which also revealed weaknesses in districts with only one use and raised awareness of the value of adaptable, easily accessible, and environmentally friendly public areas (Cheshmehzangi 2020). Because to its excellent transit links, variety of uses outside of workplaces, and capacity to accommodate social contact outside of the nine-to-five routine, Collins Square has demonstrated resilience. However, the precinct's ability to adapt to accommodate hybrid working, greater residential integration, and more sustainable mobility will determine its long-term viability.

Lessons in inclusion and equity are also included. Expensive business districts frequently run the risk of establishing enclaves, which are areas that are seen as exclusive or cut off from the city's social variety (Cheshmehzangi 2023). By planning for openness, connecting to public transportation, and creating areas that cater to a diverse range of users, Collins Square has attempted to counteract this. However, more may be done to incorporate community partnerships, inexpensive options, and design concepts that represent larger social narratives into the development fabric. Future vertical precincts' placemaking needs to take living diversity and architectural polish into consideration.

Collins Square suggests a feasible, albeit incomplete, future for Australian communities hoping to grow or enliven their central business districts. It demonstrates that vertical design can nevertheless foster human scale, movement, and encounter and that commercial development does not have to be lifeless or isolated. It affirms that the emergence of true urbanity requires interaction with context, history, transportation, and daily life. However, it also indicates that the planning and design of vertical commons must continue to examine, modify, and include broader conceptions of public value.

Lastly, Collins Square is a developing platform rather than a final answer; it has established solid groundwork for urban intensity and vibrancy (Biddulph 2011) but still has to adjust to the city's shifting demands and identities (Jayne and Bell 2017). Its teachings are found in both its accomplishments and its unembraced potential. The challenge for the next generation of mixed-use precincts is clear: think carefully, build high, and plan for the whole range of urban activities.

References

Biddulph M (2011) Urban design, regeneration and the entrepreneurial city. Prog Plann 76(2):63–103

Cheshmehzangi A (2020) 10 adaptive measures for public places to face the COVID 19 pandemic outbreak. City Soc (Washington, DC) 32(2):10–1111

Cheshmehzangi A (2023) City life experiences: lessons learnt for future urban regeneration strategies. In: Mapping urban regeneration: city life experiences in Yunnan, China. Springer Nature Singapore, Singapore, pp. 239–246

Davison G (1998) Images of modern Melbourne, 1945–1970. J Aust Stud 22(57):145–161

Jayne M, Bell D (2017) City of quarters: urban villages in the contemporary city. Routledge

Magdy Shafiq Mohamed K (2022) The Melbourne journey towards improved city liveability: a revival of a dead city (case of Melbourne CBD). Eng Res J 176:1–15

Whittle K (2016) A case study on designing super slim in Melbourne. In: CTBUH research paper, structural engineering–materials, performance and techniques, pp 1208–1215

Zakariya K, Harun NZ, Mansor M (2016) Space and sociability: mapping Melbourne's city square. Asian J Qual Life 1(2):45–55

Chapter 9
Smart from the Start: Maroochydore City Centre and the Future of Urban Integration

9.1 Foundations of a New Urban Core: Reimagining Regional Urbanism

In remote Australia, the Maroochydore City Centre is a trailblazing example of modern urban development. The new core of Maroochydore is a masterplanned project that was intended to incorporate adaptability, creativity, and liveability from the beginning, in contrast to typical city centers that change over time. By integrating sustainable practices, smart technologies, and community-focused architecture into its fundamental foundation (Daffara 2011), this ambitious project completely reimagines what a central business district (CBD) is.

The Maroochydore City Centre, which occupies 53 hectares, is a daring reinvention of what a regional CBD may be, rather than just an extension of pre-existing urban regions. With the use of cutting-edge infrastructure and careful urban planning, the project's ambition integrates commercial, residential, and recreational areas. This coordinated strategy guarantees that the city center serves as a thriving communal area that meets a range of requirements in addition to being a center for economic activity.

The Maroochydore City Centre's focus on intelligent integration is a crucial component of its growth (Gould 2005). Modern technologies have been integrated into the project from the beginning to improve sustainability and efficiency. The city center is positioned as a paradigm for future urban developments because to features including automated waste collection systems, high-speed internet connectivity, and energy-efficient structures. These developments not only raise the standard of living for locals and employees, but they also draw in companies looking for a contemporary, flexible setting.

Additionally, connectivity—both within the precinct and to the larger region—is given top priority in the city's architecture. Accessibility and easy mobility are guaranteed by integrated public transportation alternatives, pedestrian-friendly walkways, and well-placed amenities. In line with current urban living trends, this emphasis

A. Cheshmehzangi, *Australia's Urbanity*, Urban Sustainability,
https://doi.org/10.1007/978-981-95-3881-2_9

on connectivity promotes active lifestyles and a sense of community. Furthermore, placemaking is a key component of the development strategy for the Maroochydore City Centre. Public areas are carefully planned to promote community involvement, cultural expression, and social interaction. In order to create a vibrant and welcoming urban environment, green areas, public art displays, and community gatherings are essential elements. This strategy not only improves the city center's visual appeal but also fosters civic pride and social cohesiveness.

In conclusion, the Maroochydore City Centre is a prime example of a progressive approach to regional urban development. It establishes a new standard for CBD development by fusing sustainable methods, smart technologies, and community-focused architecture. This study shows that regional cities may develop thriving, liveable, and resilient urban cores that satisfy the changing requirements of their citizens via strategic planning and creative thinking.

9.1.1 From Aerodrome to Urban Engine: Strategic Positioning and Planning Vision

Strategic vision and cooperative planning are responsible for the Maroochydore City Centre's development from an abandoned airport site into a bustling urban center. The Sunshine Coast Council started an ambitious project to transform the unused property into a vibrant city center that would act as the hub of the area after realizing the necessity for a consolidated administrative and economic district.

Population expansion, economic diversification, and the need to boost regional competitiveness were some of the factors that led to the decision to create the Maroochydore City Centre. The council wanted to draw in investment, create jobs, and serve as a hub for community events by developing a purpose-built central business district. The goal of this calculated placement was to strike a balance between the needs of contemporary urban life and the area's natural resources.

The creation of a governance system that promoted effective decision-making and stakeholder participation was a crucial part of the planning goal. The Sunshine Coast Council established SunCentral Maroochydore Pty Ltd (Sunshine Coast Council, n.d.) as a wholly-owned company to provide a specialized organization to manage the project's planning and execution (Sunshine Coast Council 2015). This structure enabled streamlined processes and ensured that the development remained aligned with the overarching vision.

Transparency and community involvement were also prioritized during the development phase. To make sure the development represented the goals and requirements of the local community, extensive discussions were held to get feedback from businesses, residents, and other stakeholders. The project's continued success has been greatly aided by the sense of ownership and support that this inclusive approach has generated.

In terms of urban design, the Maroochydore City Centre masterplan was created to encourage resilience, adaptation, and sustainability. Green areas, mixed-use zones, and infrastructure that enables future development and technology developments are all incorporated into the layout. The city center is prepared to adapt to shifting sociological trends and economic conditions by placing a high priority on flexibility.

Additionally, the development vision's incorporation of smart city principles highlights the dedication to innovation. Proactive management of urban resources is made possible by the deployment of digital infrastructure, such as data analytics tools and high-speed broadband, which improves service delivery. The Maroochydore City Centre is positioned as a leader in using technology to enhance urban living because to these characteristics. The Maroochydore City Centre's strategic location and thorough planning have created a strong basis for its development into a thriving and sustainable urban centre. The project is a prime example of how regional centers may effectively negotiate the challenges of contemporary urban development through innovative thinking, cooperative governance, and creative leadership.

9.2 Built-In Intelligence: Infrastructure, Design Logic, and Smart Urban Fabric

The Maroochydore City Centre is a prime example of a progressive approach to urban development, where responsive design and digital infrastructure come together to produce a sustainable and flexible urban setting. This master-planned zone sets a new benchmark for Australian city-building by combining flexible utilities and smart technologies.

Maroochydore's extensive smart infrastructure lies at the core of its innovation. A high-speed fiber-optic network in the city center offers dependable and quick internet access to both businesses and households. Numerous smart city applications, including as public Wi-Fi, environmental monitoring, and real-time traffic management, are supported by this digital backbone. By integrating these technology, the city's services become more efficient and its residents' quality of life increases.

Maroochydore's Automated Waste Collection System (AWCS), the first of its kind in Australia, is a notable aspect of the city's infrastructure. Traditional garbage trucks are no longer necessary thanks to this underground vacuum system, which moves rubbish via a 6.5 km network of pipes at up to 70 km/h. The AWCS eliminates greenhouse gas emissions related to waste collection in addition to reducing traffic congestion and noise pollution. The method helps create a cleaner and more sustainable urban environment by simplifying waste management.

The incorporation of green spaces within Maroochydore's urban fabric is emphasized in its design. In addition to supporting biodiversity, the city center's continuous green lanes give locals easy access to recreational spaces. To promote harmony between the natural world and the built environment, these corridors are positioned thoughtfully to link parks, waterways, and other natural features. By enhancing air quality and controlling stormwater runoff, green infrastructure not only improves the city's aesthetic appeal but also supports environmental sustainability.

A modular approach to urban development is used in the city's blueprint, enabling flexibility and adaptability over time. The city can adapt to evolving demands and technological advancements with little disturbances thanks to this planned approach. Scalability is considered in the design of buildings and public areas to guarantee that the city can support future expansion and innovation. Because of its innovative approach to urban development, Maroochydore is positioned as a resilient and future-ready city.

Maroochydore's commitment to sustainability is seen in its construction methods. In order to lower energy usage, the city center uses energy-efficient designs that make use of natural lighting and ventilation. Buildings are designed to meet strict environmental performance criteria and are built using sustainable materials. These practices not only reduce the environmental impact of construction but also promote long-term cost savings and healthier living environments for residents.

The integration of sustainable design and smart infrastructure in urban development is exemplified by Maroochydore City Centre. Its creative approaches to modular planning, waste management, integrating green spaces, and sustainable building techniques show how technology and design can be used to create resilient and liveable communities. As urban areas continue to evolve, Maroochydore stands as a testament to the possibilities of thoughtful and intelligent city planning.

9.3 Animating the Grid: Public Realm, Mixed Use, and Everyday Urban Life

With careful planning to support a wide range of activities, flows, and experiences, the Maroochydore City Centre is developing into a thriving urban zone. This master-planned complex combines public areas, civic structures, residential neighborhoods, business districts, and organized events to create a vibrant setting that embodies contemporary urban living. The precinct's architecture places a strong focus on activating the public domain. Accessible green areas that promote outdoor recreation and social interaction are made available to locals and tourists alike through expansive, beautifully manicured boulevards and connected parklands. In addition to being visually beautiful, these spaces also host local gatherings, marketplaces, and cultural festivals, which promote a feeling of community and involvement.

The city center's function as a gathering place is further strengthened by the incorporation of civic buildings there. To encourage diversity and civic engagement, facilities like libraries, community centers, and cultural institutions are positioned to be widely accessible. These organizations make sure that the city center is a hub for community life by providing a variety of services and programs that appeal to a wide range of demographic groups.

A variety of housing alternatives that accommodate different lifestyles and income levels define residential development in Maroochydore City Centre. The housing stock, which includes townhouses and high-rise flats, is made to suit seniors, families, and single people. This housing diversity supports the precinct's objective of being a liveable and inclusive urban environment and encourages a balanced community makeup. The city center's commercial clusters are carefully designed to promote job development and economic vibrancy. The precinct is dotted with office buildings, shops, and hospitality establishments, promoting a smooth transition between work and play. By guaranteeing a constant stream of people throughout the day and into the evening, this integration not only strengthens the local economy but also adds to the precinct's vibrancy.

Placemaking techniques are essential for bringing life to the grid of the city center. A feeling of place and identity is enhanced by interactive digital signage (Cheshmehzangi 2022), well-designed street furniture, and public art displays. Together with dynamic programming like live concerts and community workshops, these components produce an engrossing urban environment that encourages involvement and exploration. Connectivity and walkability are also given top priority in the architecture of Maroochydore City Centre. Throughout the precinct, there are bike lanes and pedestrian-friendly walkways, encouraging environmentally friendly forms of mobility and lowering dependency on automobiles. In addition to improving environmental sustainability, this strategy promotes healthy living among locals and visitors/tourists.

To sum up, the Maroochydore City Centre is a prime example of an all-encompassing approach to urban development, where a vibrant and welcoming urban environment is produced by the interaction of public areas, civic facilities, residential diversity, business vibrancy, and placemaking techniques. The precinct successfully animates its grid through careful planning and design, providing a model for future urban centers that aim to strike a balance between community involvement and utility.

9.4 Lessons from a Blank Slate: Precinct-Making, Regional Urban Identity, and the Long Game

Building a city centre from scratch is both a radical opportunity and a deep challenge. Unlike infill developments or brownfield regenerations, the creation of Maroochydore City Centre presented a rare moment in urban planning history: the chance to shape not only form and function but also to project a new identity, cultural ethos, and urban rhythm for an entire region. In this context, Maroochydore's emerging city core becomes more than a development project—it is a laboratory for testing the long game of Australian precinct-making in the twenty-first century.

The goal of the Maroochydore City Centre was not just to increase infrastructure or unleash development potential from the beginning. The Sunshine Coast, a region long characterized by scattered townships, coastal expansion, and a car-dominant culture, needed a central location. How to create a cohesive urban identity in the absence of a solid precedent was a challenging topic that the project had to address (Cheshmehzangi 2020). The solution was to turn inward with placemaking techniques that promote public belonging and community connection while also looking forward with smart city frameworks, sustainable architecture, and future-ready infrastructure.

Throughout the precinct, there is a noticeable conflict between innovation and authenticity. On the one hand, there is the sleek, new development: grids of streets with efficient and readable layouts, digital infrastructure integrated into the design, sustainable and modular building techniques, and managed land release systems. However, there is a conscious effort to find one's identity, which is anchored in the local people, culture, geography, and memory. The city center must create its own pulse in the absence of inherited structures or long-standing urban customs (Wang and Fouseki 2025). The long game really starts to play out here. Identity, particularly urban identity, is not created at the time of opening but rather develops over time by regular interactions, events, challenges, and transformations.

Maroochydore has taken a measured approach to tackling this identity-building task. Public art, event venues, and civic structures have all been planned as symbolic anchors in addition to having practical uses. Markets, performances, and seasonal events are examples of programming that offers early support for a feeling of community. The precinct's open layout provides flexibility, allowing the areas to change as the city and its inhabitants becoming more diverse. It is consciously acknowledged that the precinct needs to accommodate unforeseen applications, unplanned events, and cultural layering—elements that are fundamental to the site yet cannot be completely dictated by a design.

However, beginning from scratch has limitations and hazards. Early activation may be more reliant on programming and curation when there are no pre-existing communities. Sometimes the coherence of vision runs the risk of monotony or a lack of grit, which is the very texture that lends older precincts their authenticity and character. It's always a negotiation to strike a balance between short-term occupancy requirements and long-term flexibility. In this way, the development of Maroochydore's City Centre will involve social processes and governance just as much as it will involve physical infrastructure (Russo 2015). Cultural sensitivity, support for local business, participatory decision-making, and a readiness to experiment and adjust are all necessary.

The project's geographical context is one of its most important contributions to Australia's urban future. Maroochydore is not a global metropolis or a capital city. It is a regional hub that has been influenced by low-rise, low-density development, population variety, and scenic surroundings. It makes a strong statement to present such a location as an example of intelligent, sustainable, and people-centered urbanism. It implies that ambitious city-making is not only a Sydney, Melbourne, or Brisbane specialty. It gives regional Australia a fresh sense of self-assurance and visibility by decentralizing the notion of what an urban future might look like.

The history of Maroochydore is constantly being written (Alcorn 1992). Over the next ten years, many of its most intriguing dynamics—such as those pertaining to housing affordability, cultural production, Indigenous participation, youth migration, and climate resilience—will only become more apparent. However, the initiative provides definite insights even at this early level. It demonstrates that precinct creation needs to be motivated by more than just economic considerations. It demonstrates that, when done carefully and modestly, design may help to shape regional identity. It also serves as a reminder that a city's prosperity is determined by the diversity of daily life that takes place there rather than by the tallest structures or the newest technology.

Maroochydore City Centre is ultimately a wager on the future—a future in which smaller cities assume leadership positions, where creativity and genuineness coexist, and where the concept of a "CBD" is transformed from a site of exclusion or spectacle to one of inclusion, interaction, and significance. That's the promise—and the challenge—of starting from scratch when creating a metropolis. This promise is weighted with optimism. It challenges communities, planners, designers, and legislators to reevaluate scale, agency, and purpose. Maroochydore poses the question, "What if urban centers were constructed to serve the rhythms of everyday life from the outset?" rather than copying the designs of capital cities. What if a city's default stance was to seize opportunities rather than respond to legacy constraints? Land use and zoning are only one aspect of this; other aspects include fostering culture, encouraging experimentation, and creating environments that support a variety of futures.

Therefore, it is the duty to make sure that the precinct becomes a genuine conduit for lived experience rather than a polished shell. Its success will be determined by how meaningfully the land is inhabited, not by how rapidly it is filled. And in doing so, Maroochydore might be composing the next chapter in the urban imagination of Australia.

References

Berenis Cecile A (1992) Maroochy towns: a study of factors contributing to the formation and growth of towns in a Queensland district. M.A. Thesis, School of History, Philosophy, Religion, and Classics, The University of Queensland. https://doi.org/10.14264/189944

Cheshmehzangi A (2020) Reflecting on Urban identities. In: identity of cities and city of identities, pp 133–178. Singapore, Springer

Cheshmehzangi A (2022) The impacts of ICT platforms on public place use in China. Built Environment 48(2):264–279

Daffara P (2011) Rethinking tomorrow's cities: emerging issues on city foresight. Futures 43(7):680–689

Gould S (2005) Maroochy 2025 community visioning & action-A case study of anticipatory action learning practices in use within Maroochy Shire, Queensland Australia. J Futur Stud 10(1):125–132

Russo C (2015) A critical analysis of four South East Queensland city futures initiatives. J Futur Stud 19(3):29–48

Sunshine Coast Council (n.d.) About SunCentral. https://www.suncentralmaroochydore.com.au/about/about-suncentral/

Sunshine Coast Council (2015) Project overview. https://www.sunshinecoast.qld.gov.au/council/planning-and-projects/major-regional-projects/maroochydore-city-centre-project/project-overview

Wang K, Fouseki K (2025) Sustaining the fabric of time: urban heritage, time rupture, and sustainable development. Land 14(1):193

Chapter 10
Brisbane Live: Staging the Future City by Redefining Urban Energy Through Culture, Connectivity, and Civic Imagination

10.1 From Transit Edge to Cultural Core: Rethinking Brisbane's Inner Heart

There is more to the Brisbane Live Precinct than merely a brand-new entertainment area. It represents a turning point in Brisbane's development as a city prepared to turn its neglected, infrastructure-constrained outskirts into hubs for urban cohesion and artistic expression. The area is a reaction to decades of spatial fragmentation and is situated above the Roma Street railway corridor, which has historically been dominated by transportation infrastructure and disjointed land uses. By doing this, it reflects an increasing desire in Australian communities to rethink key urban land for experience, connection, and civic pride rather than just for movement or consumption.

Roma Street Station and the rail lands around it have provided Brisbane with functional but meaningless service for many years (Builder Petrie, n.d.). Despite being one of the city's busiest transport interchanges, the area was essential for logistical mobility but lacked a stimulating social or spatial atmosphere. It remained a sort of infrastructure gap at the western side of the city, physically separated from the surrounding areas, which included South Bank, the Queen Street Mall, Caxton Street, and the cultural institutions along the Kurilpa axis (Heywood 2018). By placing civic aspirations on top of a functional transit corridor, the Brisbane Live plan bridges this gap. It aims to develop atop infrastructure, both physically and figuratively, through public space, architecture, and a new cultural logic for urban vibrancy, as opposed to moving or burying it.

Brisbane Live is well positioned to support long-term objectives for population expansion, economic resilience, and global positioning, all of which are in line with Brisbane's larger strategy for inner-city intensification and diversification. This precinct is not developing alone; rather, it is a part of a deliberate effort to enhance Brisbane's standing as a liveable, culturally expressive, and globally competitive city (Cheshmehzangi 2020). This precinct proposal's timing is significant, given the 2032 Olympic and Paralympic Games are approaching and a record-breaking

A. Cheshmehzangi, *Australia's Urbanity*, Urban Sustainability,
https://doi.org/10.1007/978-981-95-3881-2_10

infrastructure investment is currently under way. It acts as a practical and symbolic gesture, a pledge that the city of the future would be open, vibrant, and architecturally ambitious.

Brisbane Live's ability to bridge divides is what makes it so important to the city's urban story. With expanded and increased permeability to nearby places including the Kurilpa Cultural Precinct, Caxton Street's dining spine, and eventually the future Cross River Rail initiatives, it seeks to reestablish the western boundary of the CBD with the revitalized Roma Street Parklands (Miller and Kennedy 2012). Restoring the civic experience to the forefront of the urban agenda and re-stitching areas of the city that have long felt neglected or isolated are two aspects of this emotional and spatial reconnection.

A larger cultural goal is reflected in the decision to anchor the area with a big arena that can accommodate events, concerts, and large-scale performances. Brisbane Live is about creating an expressive identity for Brisbane as a city of events, culture, and public life, not only about density or infrastructure. The arena serves as the focal point for the planning of restaurants, retail establishments, office buildings, and pedestrian traffic (Bajracharya and Khan 2019). Crucially, the arena is positioned in the middle of the city rather than tucked away on the outskirts, confidently expressing the importance of culture in day-to-day life.

The concept of elevation, both in terms of actual height and in terms of aspiration, is equally significant. The design opposes the conventional division of movement and meaning by situating a whole precinct atop an active transit corridor. Rather, it incorporates various levels into a complicated yet approachable landscape of public life. This new verticality reflects Brisbane's evolving urban profile, which is moving away from flatland central business district patterns and toward layered urbanism where commerce, culture, and transit coexist.

Importantly, Brisbane Live does not seek to mimic the crowdedness or density of other cities across the world. Instead, it aims to provide an open, porous, climate-responsive, and socially inclusive urban lifestyle that is exclusive to Queensland. The focus is on designing areas that permit pause and interaction, readable circulation networks, and shaded walkways. It reimagines the city square and event plaza for subtropical settings, eschewing the need for a closed-off or highly structured layout in favour of promoting interaction among various user groups and at various times of the day.

The Brisbane Live Precinct provides a physical and metaphorical platform for the city's future as its urban center becomes denser and more diverse. It is a concept for urban restoration as well as architecture and land use. It turns a long-neglected transportation fringe into a cultural center where people, ideas, and events can reshape identity and provide life to the area. It does this by essentially laying the groundwork for a new performance of what inner-city Brisbane can become.

10.1.1 Urban Repair and Opportunity: The Transit-To-Place Transformation

An obvious example of urban repair is the transformation of the Roma Street transport corridor into the Brisbane Live Precinct. This act of spatial healing recognizes the effects of infrastructure-led fragmentation and aims to restore the site as a dynamic and integrated component of the city's urban fabric. Despite being essential to its operation, this area of Brisbane's city center was physically and aesthetically isolated for many years. At Roma Street, heavy rail, busways, elevated highways, and service infrastructure came together to form a sort of infrastructure canyon that divided nearby precincts and diminished the civic potential of the location. By converting what was formerly a transitional area into a vital hub of cultural activity and public movement, the Brisbane Live proposal offers a purposeful counterbalance to this history.

A strategic appreciation of the importance of centrality is at the core of this change. Roma Street holds a special place in Brisbane's inner city, both geographically and symbolically. Located at the intersection of several axes—commercial, cultural, recreational, and transportation—it marks the western edge of the central business district. It connects the Queen Street Mall spine to the parklands; it borders South Brisbane and Spring Hill; and it lies adjacent to Caxton Street, Suncorp Stadium, and the Kurilpa precinct. However, transit systems' supremacy overshadowed this powerful spatial possibility for the majority of its contemporary history. By reaffirming the site's civic function and incorporating uses that welcome, entertain, and connect, Brisbane Live's planning logic aims to unleash this latent potential.

The overarching strategic goal of making the city center more resilient, walkable, and cohesive has been one of the main forces behind this change. Because of Brisbane's continuous transition to polycentric growth and inner-city intensification, every element of the central area must not only perform well on its own but also have a meaningful relationship with its neighbours. This demand is immediately met by the Brisbane Live Precinct, which overlays coarse-grained infrastructure with a fine-grained urban fabric. By bringing cultural, entertainment, retail, and hospitality uses to an area that had long lacked such engagement, it fills in both programmatic and physical voids.

An essential component of this repositioning is transport integration. The Brisbane Live approach accepts and expands upon the current rail and bus infrastructure rather than seeing it as a barrier to be overcome or avoided. The precinct is situated on top of one of Brisbane's most important mobility hubs, thanks to the nearby Cross River Rail project, which will include a new Roma Street underground station and other improvements to the inner-city transport system. This degree of intermodality turns into a strength, facilitating more foot traffic, enhancing accessibility for major events (Williamson 2011), and guaranteeing that the precinct is not secluded but rather inextricably tied to the city's everyday rhythms.

Another issue that distinguishes the concept is urban permeability. In the past, the Roma Street location served as a barrier—a wide, inhospitable expanse that disrupted pedestrian pathways and did not provide logical east–west or north–south mobility. By suggesting a continuous series of public areas, raised plazas, lanes, and corridors that encourage rather than discourage movement, the Brisbane Live vision defies this reasoning. By providing opportunities for pause, interaction, and orientation, these novel spatial connections are intended for experience as much as circulation. As a result, the change goes beyond being merely aesthetic; it reworks the precinct's spatial DNA to better suit the objectives of inclusion, openness, and access.

This particular urban restoration initiative is noteworthy because it combines hard and soft infrastructure. Brisbane Live delivers social, cultural, and experiential infrastructure that promotes a more comprehensive urban identity in addition to filling up gaps in movement networks and providing new surfaces for construction. Together, event spaces, culinary precincts, digital art installations, and green public areas create a civic repertoire that fosters a sense of community, self-expression, and pride. Just as crucial as the site's practical makeover is its cultural and emotional activation.

Furthermore, this development's long-term urban objectives support the notion of a city that builds with the past while looking to the future. By incorporating legacy infrastructure into a multi-layered, future-ready urban environment, the new precinct elevates Roma Street's transportation history both literally and figuratively. This is in line with the global trend toward creating city cores that are smarter, denser, and more adaptable so they can change to meet changing demands without losing their accessibility or authenticity.

All things considered, the Brisbane Live Precinct is an example of a novel approach to urban reclamation that recognizes the complexity of legacy infrastructure while rejecting the idea that fragmentation is unavoidable. It redefines how cities might turn their most limited terrains into their most celebrated ones by presenting transportation space as a civic opportunity. Once characterized by mobility devoid of experience, Brisbane's western CBD edge is now starting to represent innovation, connection, and the resurgence of urban promise.

10.2 Architecture of Convergence: Vertical Density, Public Space, and Placemaking

The Brisbane Live Precinct is a prime example of a progressive approach to urban planning, where public space and vertical density come together to create a vibrant, people-centered area. This proposed area, which is located above the Roma Street railway corridor, would reshape Brisbane's cityscape by incorporating a large arena, residential and commercial skyscrapers, public plazas, pedestrian networks, and lifestyle programming.

The Brisbane Arena, a 17,000-seat venue at the center of the district, can accommodate a wide range of events, including concerts and indoor sports. The site's historical significance as a transportation center is reflected in its architectural shape, which is inspired by fluidity and mobility. Three structures devoted to business, lodging, and residential purposes encircle the arena; among them is a distinctive 90-story residential skyscraper that will stand out as a distinguishing element of the city's skyline. A variety of dining, shopping, and movie theatre activities are intended to be hosted on the podium along Roma Street in order to energize the street edge and foster a vibrant urban atmosphere.

The precinct's architecture places a strong emphasis on urban permeability, guaranteeing smooth linkages between various metropolitan areas. To improve accessibility and promote foot traffic, equitable pedestrian connections are being developed to connect King George Square, Wickham Park, Roma Street Station, and Platform 10. The district is positioned as a key hub in the city's transportation system because to the integration of significant public transportation facilities, such as the future Brisbane Metro, the projected Cross River Rail, and the historic Roma Street station.

The precinct's design has a strong emphasis on public space, and the existing Roma Street Parklands would be improved with an extra 12 hectares of subtropical parkland and leisure area. By offering places for socializing and rest, these green spaces hope to improve the precinct's liveability and environmental sustainability for both locals and tourists. A readable and vibrant urban environment is produced by the thoughtful planning of the precinct's mixed-use programming and architectural scale. The precinct can accommodate a wide variety of activities and user groups thanks to the combination of high-density building and accessible public areas. The precinct's cultural offerings will be further enhanced by lifestyle programming, which includes cultural facilities influenced by organic shapes and fluidity. These facilities are expected to serve as theatre and museum sites.

In conclusion, the Brisbane Live Precinct exemplifies a comprehensive approach to urban planning, combining placemaking techniques, public space, and vertical density to provide a lively and welcoming cityscape. The precinct is positioned to become a major center of activity and a template for upcoming urban developments in Brisbane and abroad by combining various activities and encouraging connection.

10.3 Performing Urban Life: Activation, Experience, and Cultural Economy

Brisbane Live is a proposal for the potential performance of the city itself, not just a development. Fundamentally, the precinct is intended to serve as a venue for collective experience, expression, and activation in addition to serving as a container for activities. Its potential is found in its capacity to orchestrate life on both a large and small scale, from international entertainment events to the everyday encounters that characterize the atmosphere of a city. By doing this, Brisbane Live establishes itself as a force for change in the city's urban identity and cultural economy.

The planned arena, a 17,000-seat space that will serve as the city's cultural amplifier, is the focal point of this performative concept. The arena is not meant to be isolated; rather, it is intended to create a dynamic flow of people before and after events by spilling activities into the surrounding precinct. The arena will act as a magnet, drawing a variety of audiences into the city's western center and dispersing energy across the streets, plazas, and business frontages for everything from sporting events and concerts to civic ceremonies and exhibitions. It reinterprets going to an event as a more comprehensive urban experience that encompasses the precinct's architecture, hospitality, and open space in addition to the performance itself.

A carefully chosen network of retail and lifestyle options that liven up the public realm's edges further supports this concept. A permeable bridge between commercial programming and the larger civic environment will be created by the integration of high-quality food, beverage, and retail establishments throughout the precinct. These are not passive amenities; rather, they are a component of a broader performative approach that aims to keep the precinct vibrant even when there are no major events taking place there. Brisbane Live is an all-day, all-season destination thanks to its laneway pubs, activated rooftops, and outdoor eating options, which offer several chances for social interaction.

Critically, its vitality is not limited to large-scale events. Smaller-scale performance events, such as buskers on terraced terraces, pop-up art exhibits on digital screens, outdoor movies in public squares, or unplanned meetings on pedestrian bridges, are supported by the precinct's master plan. The architecture makes sure that these micro-performances, which are the spontaneous manifestations of a lived city, have space to manifest. The precinct can function at different tempos—fast and slow, planned and unplanned, lonely and collective—by combining formal and informal activities.

Given Brisbane's subtropical climate, the emphasis on outdoor experiences is particularly noteworthy. In order to entice people to stay, the precinct incorporates shaded walkways, airy arcades, and raised green platforms. Instead of enclosing internal activities, Brisbane Live blurs the lines between indoor and outdoor spaces, transforming the city itself into a form of public theatre. Public areas are intended to be active stages—arenas in and of themselves for civic engagement, leisure, and cultural expression—rather than transitional places.

Another essential component of the precinct's performative logic is the economy at night. Brisbane Live aims to prolong urban activity late into the evening, even if many areas of Brisbane's inner city still lose their vibrancy after dark (Mccrea and Walters 2012). The precinct intends to de-risk evening activity and concentrates after-hours vitality in a highly walkable, safe, and planned environment by concentrating venues, hospitality, and transportation access in one location. The nighttime offering is an essential component of the precinct's cultural and economic justification, not just an add-on. In addition to meeting the lifestyle demands of locals, employees, and tourists, it advances Brisbane's goal of becoming a globally competitive event city.

Placemaking thus turns into a cultural narrative as well as an instrument for urban planning. Each surface, sightline, and sequence are carefully chosen to encourage participation, whether it be through social invitation, spatial comfort, or aesthetic pleasure. Brisbane Live turns into a district that co-produces events rather than just hosting them. It transforms structures into public life facilitators and space into experience. This performative urbanism is not about spectacle for its own sake but about embedding cultural depth and emotional resonance into the city's everyday scenes.

There are important ramifications for Brisbane's cultural economy. By connecting with Caxton Street, South Bank, and the Queen's Wharf development in a larger constellation of urban energy, Brisbane Live is positioned to boost the city's standing in the national and international entertainment circuits, draw in fresh types of creative investment, and stimulate the growth of nearby precincts. It suggests a paradigm in which cultural programming and urban design are interwoven systems rather than parallel tracks, one enhancing the other to provide life to the lived experience of place.

In sum, Brisbane Live is an intentional effort to bring urban life to life by bringing people, programs, and architecture together. The precinct was created not only for its contents but also for what it makes possible, including the civic stage it creates for Brisbane's upcoming era of urban expression, the performances it welcomes, and the energy it directs.

10.4 Beyond Spectacle: Lessons for a Next-Generation Civic Precinct

At first look, Brisbane Live appears to be a development of remarkable ambition, a mixed-use, high-density entertainment district that combines civic symbols with commercial power, cultural economy with transportation infrastructure. However, its greater potential to change how cities view civic place-making in the era of urban growth and cultural convergence is more promising than its ability to impress. Organizing more events is not the only goal of our project. It makes a statement on how to make those occurrences significant to a city's daily existence.

The fact that genuine urban vibrancy cannot be masterplanned as spectacle alone is one of the fundamental lessons Brisbane Live imparts. One major issue that high-profile precinct frequently face is how to maintain their relevance and vitality in between high-profile events. The quieter moments i.e., the lunch breaks, the late-night wanderings, the weekend rituals, are what ultimately mould public memory and civic attachment, even when a sporting event, international concert tour, or political convention may garner the most attention. To address this, Brisbane Live's design incorporates adaptability, accessibility, and layered use throughout its footprint. This makes sure that the precinct is distinguished not just by its events but also by its capacity to accommodate the rhythms of daily life.

The significance of integration serves as a second lesson. The precinct's position over the transit corridor on Roma Street is more than just symbolic. Reassembling broken pieces of Brisbane's inner core and transforming infrastructure gaps into civic ties is a deliberate act of urban repair. If successful, Brisbane Live will seamlessly integrate with Caxton Street, Roma Street Parklands, the Cross River Rail, and Brisbane Metro, overcoming the isolation that mega-precincts frequently endure. This degree of spatial integration and transportation is fundamental, not incidental. In this instance, movement and porosity are just as important to urban vitality as structures and programming.

However, great difficulty comes along with audacious ambition. The issue of authenticity is among the most important criticisms thrown at well-publicized advances of this type. Will Brisbane Live resemble a packaged district or a municipal precinct? Will its public function be overshadowed by its commercial layers? These are serious issues. Urbanity is not created solely by density and design, as cities all around the world have discovered. Although they are more difficult to create, emotional connection, social inclusion, and cultural resonance are crucial for success. Brisbane Live must so refrain from turning into a corporate haven. If it wants to be considered a true municipal anchor, its street-level activation, pricing tactics, programming diversity, and public space access must all continue to be inclusive.

Another conflict that needs constant attention is accessibility, which affects not just physical design but also economic and cultural engagement. The precinct runs the danger of alienating the very people it is trying to reach if it is dominated by upscale consumption or limited to venues that require tickets. Cultural precincts that provide everyone, regardless of age, background, or income, with moments of surprise and belonging are the finest. In this sense, Brisbane Live should strive to democratize its experience, not just in its public areas but also in its local collaborations, creative programs, and community gatherings.

Brisbane Live also pushes the topic of urban legacy into a wider discussion. It comes as Brisbane is getting ready for international notice, especially in light of the upcoming 2032 Olympic and Paralympic Games. However, long after the games are over, the true impact of a precinct like this will be evaluated. Its ability to integrate into city life, adjust to shifting populations, maintain environmental resilience, and continue to encourage participation over decades rather than just project cycles will determine its long-term worth.

This has implications for planning governance as well. Brisbane Live is a prime example of the kind of intricate, multi-stakeholder cooperation that characterizes urban initiatives in the twenty-first century, wherein community organizations, commercial developers, infrastructure operators, state agencies, and cultural institutions must work together. Although coordination is a significant obstacle, it also offers a chance to rethink city planning, emphasizing processes and experiences that support public life rather than just constructing buildings.

Overall, Brisbane Live is more than just an entertainment district. It is a test of the ability of contemporary cities to create environments that combine infrastructure and creativity, intimacy and intensity, and spectacle with substance (Cheshmehzangi et al. 2025). Its teachings are unmistakable: programming must be inclusive, networking must be smooth, identity must be developed rather than forced, and design must be generous. Only then will such a district be able to go beyond mere showmanship and turn into a long-term investment in Brisbane's civic life.

References

Bajracharya B, Khan S (2019) Urban governance in Australia: a case study of Brisbane City. In: New urban agenda in Asia-Pacific: governance for sustainable and inclusive cities. Singapore, Springer, pp 225–250

Builder Petrie J (n.d.) Roma street railway station. Coordinates, 27, 153–01914692.

Cheshmehzangi A (2020) Identity of cities and city of identities: examples. In: identity of cities and city of identities. Singapore, Springer, pp 179–244

Cheshmehzangi A, He B, Zuo T (2025) Strengthening health-resilience Nexus through sustainable urban design in Brisbane, Australia. In: designing healthy cities: integrating climate-resilient urbanism for sustainable living. Springer Nature, Singapore, pp 163–184

Heywood P (2018) Growth plans and growing pains in Brisbane's sunbelt metropolis. Urban regeneration in Australia. Routledge, New York, pp 159–177

Mccrea R, Walters P (2012) Impacts of urban consolidation on urban liveability: comparing an inner and outer suburb in Brisbane, Australia. Hous Theory Soc 29(2):190–206

Miller E, Kennedy RJ (2012) Public realm and transport, international best practice case studies and opportunities for Brisbane. Unpublished book report. https://eprints.qut.edu.au/53756/

Williamson JCR (2011) The intermodal metropolis: Spatial protocols at the convergence of regional mobility networks. Doctoral dissertation, University of Waterloo

Chapter 11
Reframing the Nation's Capital Core: The Quiet Power of Contemporary Placemaking in Civic Quarter, Canberra

11.1 A New Heart for the Capital: Urbanising the Civic Spine

Canberra, a (modern) city of politics and procedure, planned order and ceremonial geometry, is frequently viewed in the public psyche as an abstract concept rather than a lived city (Beer 2009). Its center has long been characterized more by symbolic purpose than by urban pulse, in contrast to the busy centers of Sydney or Melbourne. However, this view has begun to change in recent years. The rise of Civic Quarter, a purposeful, design-led effort to reinterpret the capital's urban character from the inside out (Deem et al. 2019), is one of the initiatives that best reflects this change.

The Civic Quarter is a gateway between the central business district and the larger urban area, located at the northern end of Northbourne Avenue, Canberra's physical and metaphorical spine. In the past, this area was more about movement than meaning; it was a boulevard of administrative tedium, with antiquated office supplies and a lackluster pedestrian experience. Even though the Griffins' initial plan for the city emphasized symmetry (Hinton 2024), clarity, and massive grandeur, daily life in the central business district frequently felt disjointed and underdeveloped. In order to anchor the city's functional growth with dignity, density, and precise architecture, Civic Quarter suggests reinterpreting that past via a modern urban perspective rather than rejecting it.

Despite not being a mega-project in scope, this precinct is significant because of that limitation. Delivering first-rate office space, a public forecourt, ground-floor activity, and integrated landscaping to a section of the city that lacked cohesion and street life before, it is a daring intervention executed with well-timed gestures. Civic Quarter provides a lesson in the value of continuity and clarity rather than producing spectacle. The goal to humanize the nation's capital, i.e., to create space for the everyday, the local, and the impromptu within the framework of national purpose, is the foundation of its urban logic.

A. Cheshmehzangi, *Australia's Urbanity*, Urban Sustainability,
https://doi.org/10.1007/978-981-95-3881-2_11

Civic Quarter prioritizes proportion, material depth, and visual transparency over height and visibility, which are the goals of many recent CBD developments (Cheshmehzangi and Tang 2024). From a distance, its structures appear austere and erect, but up close, they are textured and hospitable. Canberra's desire to reconcile its administrative symbolism with the increasing demand for everyday urbanity is reflected in this deliberate dual interpretation. Public life is encouraged here rather than required; the ground plane is viewed as a platform for interaction rather than a conduit for business.

Crucially, Civic Quarter has no intention in becoming a tourist destination. It is not a place for entertainment or novelty. Its goal is less obvious but possibly more enduring: to provide a natural and supportive environment for working, walking, dining, meeting, and loitering in central Canberra. For example, the public forecourt, which is intended for lunch breaks, casual get-togethers, and silent observation, is not merely a transitional area but rather an extension of the street life and workplace. It reflects a broader ethos in the project: that public realm is not an afterthought, but the connective tissue of contemporary urban experience.

A rising change in Canberra's definition of civic significance is also indicated by the emergence of Civic Quarter. The capital must now operate as a true city—responsive to changing work patterns, a variety of urban lifestyles, and a younger population that demands walkability, openness, and design intention—instead of the previous emphasis on institutional gravity and controlled homogeneity. Given Canberra's expanding population and diverse cultural setting, it is not only desirable but also imperative to have adaptable, high-quality urban fabric. Civic Quarter places itself in the forefront of this reorientation by offering not only infrastructure but also a model for how contemporary development may continue to be sensitive to people and place.

Strategically, the precinct also plays a vital role in shaping perceptions of Northbourne Avenue as more than a corridor. Instead of being unused space, it reclaims the avenue's margins as municipal addresses. The way the project interacts with the street is very interesting; rather than retreating behind parking lots or protective barriers, its buildings open wide to the road, interacting with passersby through active uses, seats, and planting. In a city where scale may easily become alienating and distances are great, this attention to edge conditions is essential. The architectural decisions made for Civic Quarter counteract this trend by fostering a feeling of arrival and inclusion amid the gridded formality of the city.

Civic Quarter ultimately demonstrates that urban renovation in a capital city does not have to entail erasure or spectacle. It can entail paying attention to the rhythms of everyday living and creating thoughtful, proportionate, and forward-thinking designs. In doing so, it helps transform Canberra not into something other than a capital, but into a capital that also feels like a city—lived, layered, and increasingly alive.

11.1.1 Planning, Policy, and Opportunity: Positioning Civic Quarter

The Civic Quarter precinct in Canberra was not created overnight; rather, it is the result of changing urban objectives, reevaluated policy frameworks, and a changing cultural perception of what should be available in the city center. Being the nation's capital for many years had a significant impact on the city's spatial makeup, which was organized, hierarchical, and vehicle-centric. With broad boulevards dividing precincts and low-density construction predominating even in the city's core, urban design adhered to the logic of clarity and control. Legibility and visual order were facilitated by this, but much of central Canberra also felt formal, under-activated, and occasionally cut off from the daily reality of modern urban life.

Civic Quarter represents a significant change in this regard, a reaction to the increasing realization that Canberra needs to move past the ideas of administrative function and automobile-focused infrastructure (Fischer and Weirick 2013). Over the last twenty years, the National Capital Authority and the ACT Government have progressively adopted a planning strategy that emphasizes public realm improvement, sustainable transportation, and mixed-use intensity. This strategic shift positioned the Northbourne Avenue corridor and its immediate environs as an anchor for urban renewal, paving the way for the development of Civic Quarter.

Crucially, Civic Quarter is located inside the "City Renewal Precinct" (Newton and Glackin 2018), an area that has been earmarked for human-centered development, superior design, and targeted investment (Allam and Cheshmehzangi 2024). Civic Quarter emerged as a test case for how private development could be in line with more general public goals thanks to the precinct-wide agenda. The City Renewal Authority of the ACT Government, which was set up to supervise this development, prioritized pedestrian traffic, quality design, and civic character. It provided a framework for regulations and cultural support for projects that defied the city's traditional development standards.

Long underutilized, the particular piece of property that became Civic Quarter had fragmented ownership and low-grade business stock. The site was perfect for densification because of its location at the intersection of Northbourne Avenue and the larger CBD grid, but only with careful planning. Rather than pursue piecemeal development, the approach was to craft a precinct-scale response—one that embraced verticality while giving equal attention to its interface with the public realm. In this sense, Civic Quarter was intentionally formed by a planning culture that aimed to restore the central business district as a hub for civic engagement rather than merely administrative traffic, rather than merely being allowed by law.

The creation of a new urban development typology that prioritized mid-rise density, architectural integrity, and layered programming was a significant contributor to this change. This change was mirrored in the last ten years' changes to the City Plan and Territory Plan, which mandated integration with public infrastructure, promoted openness at the grassroots level, and established incentives for precinct-scale outcomes. The realization that providing more than just workspace was essential to the city's future was reflected in these revisions. It required spaces for people to congregate and socialize, as well as amenities and readability. Civic Quarter's adept handling of this regulatory environment serves as an example of how planning can be both curatorial and facilitating, establishing boundaries while fostering creativity.

Crucially, the project also complemented Canberra's larger objectives for urban sustainability. The precinct's design decisions were influenced by the ACT Government's dedication to carbon neutrality, active transportation, and green infrastructure. Energy-efficient systems, end-of-trip amenities, and robust links to the city's light rail spine were integrated features rather than add-ons, demonstrating how policy alignment may provide observable urban results. In particular, the link to public transportation was a game-changer. The location became more accessible and visible than ever before because Northbourne Avenue became the main thoroughfare of the city's new light rail system. The justification and impetus for a denser, more people-focused urban shape came from this modal shift—from auto-dominated thoroughfares to transit-oriented development.

The time of Civic Quarter's planning was also crucial to its success. There was a growing desire for a capital city that provided more than just bureaucratic efficiency as Canberra's population grew, its economy diversified, and its culture developed. A youthful population, a flourishing education sector, and a growing services economy need a walkable, compact, and engaging city that could accommodate urban lifestyles. As a paradigm for how policy-led planning may be realized through careful, financially feasible development, the Civic Quarter precinct was positioned to meet that desire.

Civic Quarter benefited from a cultural change in the city's planning and design discourse in addition to technical policy alignment. In Canberra, civic urbanism was no longer viewed as a side issue; rather, it was now essential to the future identity of the city. The precinct's trajectory was influenced by design review panels, competitive processes, and public interaction, guaranteeing that it was not only compliant but aspirational. The end product is a project that combines place-led design, private investment, and government ambition.

In conclusion, the rise of Civic Quarter is not coincidental. Canberra's urban priorities were purposefully reevaluated, moving from being a city of movement to a city of place. The precinct is a testament to what can be achieved when land-use policy, design culture, and development goals align in pursuit of a more meaningful and human urbanism. It was made possible by updated planning frameworks, guided by urban renewal objectives, and inspired by a more civic-minded vision of the capital's future.

11.2 Architecture as Dialogue: Built Form, Landscape, and Integrated Movement

Civic Quarter's architecture exudes a calm assurance based on readability, clarity, and a deliberate conversation with its urban environment rather than grandiosity. The Civic Quarter precinct represents a modern response to Walter Burley Griffin's legacy of order, proportion, and axial clarity in a city that has long been influenced by it. It embraces the complexity of contemporary urban movement, spatial experience, and architectural expression while honouring restraint. It is a sophisticated urban choreographic exercise rather than a show of form.

The idea of integration-between structures, people, and landscape-lays the foundation of the precinct's architectural approach. In order to respond to the scale, rhythm, and tone of nearby structures and provide a more permeable and activated interface with the public realm, the Civic Quarter construction is purposefully integrated into its surrounding environment rather than existing as a stand-alone statement. The built form steers clear of sudden imposition while adopting a vertical profile suitable for Canberra's developing high-density center. Bulk is broken down, visual impact is eased, and a human scale is maintained through the use of stepping, modulation, and layered façades.

Character and continuity are established in large part through materiality. The architectural palette makes use of modern yet contextually aware stone, glass, wood, and metal finishes. Without resorting to pastiche, the buildings' use of regionally inspired textures and tones gives them a grounded aspect while recalling the capital region's geological and cultural landscape. Open ground planes, glazed façades, and double-height lobbies that encourage visual connection and indicate accessibility are some examples of how transparency is employed as a narrative technique. This openness is consistent with the precinct's civic goal of promoting transparency, trust, and involvement, especially among the commercial and governmental entities that occupy the area.

The influence that landscape design has on how people perceive location and movement is equally important (Cheshmehzangi 2014). Civic Quarter incorporates landscape as structure—a connecting thread that unifies buildings, entrances, walkways, and open spaces into a cohesive whole—instead of treating green space as ornamental or residual. Native plants, linear planting plans, and thoughtfully positioned seating areas create an environment that is both serene and urban, efficient and welcoming. The end effect is a type of spatial punctuation that strikes a balance between softness and readability rather than just landscaping. It coordinates how people move across the area while offering visual relief, promoting biodiversity, and reaffirming the precinct's environmental goals.

The Civic Quarter's movement systems are purposefully designed to encourage involvement and permeability. Desire lines are created by pathways that are straightforward but sinuous, making them seem organic rather than forced. Sightlines, illumination, and the rhythm of architectural components all contribute to the ease of wayfinding. Multi-modal access is guaranteed by the close vicinity to Canberra's light rail station on Northbourne Avenue, and there are many pedestrian walkways and bike lanes to provide alternatives to driving a car. Crucially, the design avoids the temptation to over-program or restrict, allowing for future changes in how people utilize and occupy the area. For a website that serves a variety of user groups, including office workers, public servants, students, tourists, and community members, this flexibility is essential.

The architectural shape of the precinct also adapts to the climate and time of the area. Shade structures, overhangs, and the placement of public seats and entrances all reflect Canberra's various seasons, which include scorching summers and cool winters. Water-sensitive design elements that help regulate the microclimate, such as rain gardens and permeable pavement, support the precinct's sustainability credentials. With illuminated walkways and façade lighting that highlight form without overpowering atmosphere, well-thought-out lighting strategies turn Civic Quarter into a secure and readable area at night.

The Civic Quarter's architectural dignity is among its most remarkable features. This precinct conveys confidence through proportion and discipline, in contrast to the flamboyance frequently associated with CBD towers in other cities. It is a place where design conveys value without being ostentatious, where movement is deliberate rather than rushed, and where order is not rigidity but rhythm. Rather than preventing vibrancy, this dignity places it in a context that honours both the individual's experience and the group's presence.

All things considered, Civic Quarter's architectural and spatial layout represents a type of urban sophistication sensitive to Canberra's changing requirements. It emphasizes connectedness, accessibility, and civic pride while speaking quietly and purposefully. By carefully combining landscape, constructed form, and human activity, it provides an example of how cities can modernize and densify without compromising public value, intelligibility, or character. Civic Quarter serves as an example of how architecture may be used as a dialogue tool to actively shape the quality and meaning of urban life rather than merely serving as a background.

11.3 A Working City Comes Alive: Activation, Amenity, and Civic Experience

A workplace district that not only accommodates office activity but also orchestrates a daily civic experience, Civic Quarter presents a compelling alternative in a city that is frequently characterized by its administrative formality and bureaucratic rhythm. The precinct aims to blur the lines between work and public life by combining accessibility, amenity, and human-centered architecture to create a more vibrant and welcoming urban core, as opposed to replicating the tedium of traditional commercial districts.

Deliberate dedication to ground-level activation is at the core of this change. More than just economic viability was considered when designing Civic Quarter's Street interface. In addition to foot circulation, retail frontages are designed for diversity and public interaction. Restaurants and wellness centers face plazas with plenty of outdoor seating, while cafés overflow into beautifully manicured forecourts. The design promotes visibility and engagement by encouraging these uses to function with openness and permeability, substituting calls to dwell, connect, and pause for the blank walls and idle nooks of earlier corporate spaces. The feeling of openness is the result of a design process that values civic giving; it is not an accidental effect.

The incorporation of green space and public amenities at all precinct levels contributes to this liveliness. Landscape interventions are essential structural components that encourage use all day long, not ornamental afterthoughts. Intimate pockets of meeting space are framed by soft landscaping, while tree canopies provide shade in the summer and define pedestrian pathways. In addition to providing respite from the hardscape, these green insertions serve as a reminder to stay. The rooms are proportioned and calibrated to support daily rhythms without dictating behaviour, whether it's a group of public officials having lunch on a low wall or guests hopping between meetings and public events.

Civic Quarter differs from previous Canberra business areas in that it recognizes that vibrancy must be curated through lived experience rather being produced only through density. A subtle but crucial part of animating the website is programming. As part of a larger placemaking strategy, temporary installations, wellness activations, seasonal markets, and cultural pop-ups have all been envisioned. Instead of merely serving as a corridor for business, these interventions turn the precinct into a venue for social interaction. Crucially, they also mirror the changing face of Canberra, a city that is becoming more and more populated by knowledge workers, creatives, and students who anticipate that their places of employment will also serve as experiential spaces.

This dynamic atmosphere is facilitated by the precinct's architecture, which offers adaptable and unrestricted transitional areas. Lobbies are seen as public spaces rather than private entrances. Public passageways blur the line between circulation and congregation. This blurring, which encourages individuals to inhabit the city's in-between spaces in unplanned, informal ways, is essential to the precinct's mood. Through this layering of uses, Civic Quarter starts to change how people view the heart of the capital: as a site of movement, interaction, and vitality rather than as a static monument to order.

Another crucial component of the precinct's activation is the presence of art, design, and civic narrative. Passersby are engaged by murals, sculpture installations, and interpretative features, which further emphasize the idea that this is a place of place and identity rather than just governance or business. These little bursts of visual appeal provide richness, memorability, and a human scale to the city experience without being intrusive. They provide opportunities for inquiry and conversation, two things that are frequently lacking in conventional office settings.

Civic Quarter's reputation as a center of midday energy is further strengthened by the amenities it offers. End-of-trip amenities encourage a change in mode of transportation to walking and bicycling. Easy access is given priority by utilizing the nearby transportation infrastructure. Accessibility for all users is guaranteed via inclusive design. These characteristics point to a change in focus from creating offices that isolate people to creating engaging urban environments. In this sense, Civic Quarter is a modern example of civic-commercial integration, where public life and employment are mutually supportive rather than antagonistic. Civic Quarter is a prime example of how high-functioning work areas may also serve as civic catalysts as Canberra continues to diversify. It portrays a working metropolis as one that pulsates with daily activity, interpersonal relationships, and cultural vitality rather than one that clocks in and out. The civic spine of the capital is evolving into something truly urban rather than merely administrative.

11.4 Designing with Intention: Human-Scaled Growth and Urban Identity in the Capital

Civic Quarter is a subdued yet impactful addition to Canberra's changing urban identity story. The precinct represents a change toward a more grounded and human-scaled urbanism in a city that has historically been characterized by axial grandeur, low-rise order, and administrative purpose. It reframes the symbolism that has created Canberra's form via the everyday experiences of those who live, work, and travel through the city center, without abandoning it. It shows that the civic center of the capital may be a lived, readable, and welcoming place rather than just an abstract location of authority.

The success of Civic Quarter is largely due to its obvious intentionality. It was never intended for the precinct to be iconic in the conventional sense. Its subtlety—the way it puts utility, comfort, and dignity without spectacle first—is what gives it its ambition. It reclaims the concept of the central business district as a civic forum by designing buildings and public areas that encourage habitation rather than observation. This is not the performative density of commercial downtowns or the monumentalism of national architecture. It is a paradigm of restrained perfection that affirms everyday urban belonging through design.

One of the most important contributions of Civic Quarter is its redefinition of growth. The precinct provides a vertical but finely grained extension of the city center, which is an alternative in a planning culture that is frequently fixated on scale and skyline. Its ground-level permeability, layered landscapes, and human-scaled interfaces cast doubt on the idea that intimacy must be sacrificed for growth. Rather, Civic Quarter shows that urban planning may be both efficient and compassionate—that compactness and caring can coexist.

This is especially relevant in Canberra, where the formal planning history of the city can occasionally stifle creativity. Civic Quarter softens the edges of its rigidity rather than denying it. It adds ambiguity to areas that have historically been very prescribed, promoting mobility without prescribing routes and facilitating interactions without imposing spectacle. These characteristics are essential for creating a more welcoming urban culture, particularly in a city that has to balance its roles as a developing regional hub and the nation's capital.

However, there are conflicts with the project. Civic Quarter is part of a broader system of institutional governance and market forces, and Canberra is still undergoing change. The difficulty is in making sure that its advantages—walkability, activation, and diversity—are spread throughout the urban fabric rather than being limited to a single development piece. This calls for civic leadership, policy consistency, and public–private cooperation that upholds the precinct's fundamental principles across time. The public intent of the precinct must be actively protected from the danger of being fragmented or commercialized.

Furthermore, concerns about cultural representation still exist. Deeply ingrained Indigenous history, complex immigration narratives, and developing youth cultures can all be found in Canberra. Even though municipal Quarter has made strides toward establishing a more democratic and open municipal setting, more needs to be done to guarantee that these areas represent the whole range of urban identity. If the precinct is to have a significant impact on its various publics, inclusive storytelling, collaboratively created cultural programming, and consideration for symbolic narratives must all continue to guide its development.

Finally, Civic Quarter provides a collection of urban teachings based on transformation by character rather than transformation by scale. It teaches that a city's essence can be shaped by modern civic areas that do not have to take up much of the skyline. It demonstrates how cities, particularly capital cities, can develop with humility and accuracy through well-considered architecture, balanced public life, and a subtle yet intentional interaction with Canberra's core. The real legacy of the precinct might not be found in its structures but rather in its mindset: a dedication to thoughtful design that prioritizes users and a larger urban identity. Civic Quarter in Canberra is more than just a business district or office center; it represents a new standard for how a planned city may regain its soul by deliberate and human-centered urban design.

References

Allam Z, Cheshmehzangi A (2024) The social movements and political landscape of sustainable futures. In sustainable futures and green new deals. Springer Nature Switzerland, Cham, pp 81–104

Beer C (2009) Spectacle, urban governance and the politics of nationhood: Canberra and the production of anzac day commemorations and Australia day live. Urban Policy Res 27(1):59–72

Cheshmehzangi A (2014) Spatial syntagma and identity of a place: sensing, relating to, and knowing a place. J Hum Behav Soc Environ 24(7):799–810

Cheshmehzangi A, Tang T (2024) Beijing: balancing tradition and modernisation in a metropolis with multiple agenda on sustainability and innovation. In: 30 years of Urban change in China's 10 core cities. Springer Nature, Singapore, pp 27–43

Deem J, Tiernan A, Evans M, Grattan M, McCaffrie B (2019) Beyond the Canberra Bubble. From Turnbull to Morrison: understanding the trust divide, 1–13.

Fischer KF, Weirick J (2013, November) Canberra 2013: planning and urban development challenges at the centenary of the national capital. State of Australia's cities conference
Hinton E (2024) Of other plans: re-viewing Canberra through the lens of other spaces. https://researchoutput.csu.edu.au/en/publications/of-other-plans-re-viewing-canberra-through-the-lens-of-other-spac
Newton P, Glackin S (2018) Greyfield regeneration: a precinct approach for urban renewal in the established suburbs of Australia's cities. In Urban Regeneration in Australia, Routledge, pp 249–267

Part III
Sustainable and Master-Planned Communities

Developments designed for long-term sustainability, resilience, and liveability:

- **Ecco Ripley, Ipswich**—A green-focused master-planned community.
- **Riverlea Park, Adelaide**—A large-scale residential and commercial project with community-centered planning.
- **Hills Showground Station Precinct, Castle Hill**—A transit-oriented development improving urban mobility.
- **Yarrabilba, Queensland**—A master-planned satellite city balancing urban expansion with green infrastructure.
- **Nightcliff Foreshore Revitalisation, Darwin**—A coastal renewal project enhancing public access and ecological resilience.

Chapter 12
Ecco Ripley as the New Urban Ideal in Southeast Queensland: Designing Tomorrow's Community Today

12.1 Grounded Futures: Ecco Ripley and the Reimagination of Growth

In the context of Australian urban development, Ecco Ripley stands out as a paradigm-shifting example. This master-planned community, which is located in Ipswich's Ripley Valley, addresses the urgent issues of fast population increase, climate change adaptation (Salas 2017), and changing suburban living standards. With plans to house more than 120,000 people in over 50,000 residences, Ecco Ripley is more than just a housing development; it's a comprehensive strategy for sustainable urban planning.

The genesis of Ecco Ripley aligns with the South East Queensland Regional Plan's objectives, which identified Ripley Valley as a key area for urban expansion. By designating the region as an Urban Development Area in 2010, the Queensland Government recognized the need for a new paradigm in community development and paved the way for creative planning and design. The project's developer, Sekisui House Australia, seized the chance to create a community that balances contemporary living with environmental stewardship.

At the heart of Ecco Integrating green areas with sustainable infrastructure is part of Ripley's mission. Pebbles Park and Faye Carr Park are two of the community's more than 76 hectares of parklands, which are important ecological and recreational resources. These areas support biodiversity, stormwater management, and the general well-being of the community; they are not merely decorative elements. A dedication to building a resilient ecosystem that can adjust to changing climate conditions is shown in the emphasis on green infrastructure.

Ecco Ripley's design philosophy takes social and economic factors into account in addition to environmental ones. An example of this integrated approach is the $1.5 billion Ripley Town Centre, which is located in the center of the community. It is envisioned as a mixed-use hub that blends civic, commercial, retail, and residential areas to create a thriving urban center that promotes social interaction and local employment. By minimizing reliance on private vehicles and fostering community

A. Cheshmehzangi, *Australia's Urbanity*, Urban Sustainability,
https://doi.org/10.1007/978-981-95-3881-2_12

solidarity, this centralization of resources is consistent with the ideas of walkable neighborhoods.

Ecco Ripley's transportation planning demonstrates its dedication to sustainability and connectedness. The planned extension of the Springfield railway line will help the town by providing easy access to Ipswich, Brisbane, and other locations. In the meanwhile, an active transportation strategy encourages citizens to choose more sustainable and healthful forms of transportation by supporting bike lanes and pedestrian walkways. This multimodal strategy guarantees effective and ecologically conscious mobility both inside and outside the community.

Being the first in Queensland to receive a 5-Star Green Star (Bond 2010), Communities designation from the Green Building Council of Australia is a defining feature of Ecco Ripley's development (Dharmarathna et al. 2025). The project's quality in governance, design, liveability, economic prosperity, and environmental responsibility is recognized by this ranking. It is evidence of the project's all-encompassing approach to sustainability, which includes long-term profitability and community involvement in addition to physical infrastructure.

Ecco Ripley's diverse housing options meet the changing demands of modern society. The neighbourhood provides a variety of housing options, ranging from family houses to reasonably priced apartments, guaranteeing accessibility and inclusivity. A diverse population is supported by this diversity, which creates a vibrant and strong social fabric. Additionally, the community is better positioned to take part in the digital economy thanks to the integration of high-speed internet infrastructure, which increases chances for remote work and online interaction.

The goal of Ecco Ripley's community development programs is to foster a strong sense of place and belonging (Sekisui House, n.d.). This goal is largely achieved through community initiatives, regular events, and the construction of social infrastructure like schools and medical facilities. By encouraging social capital and civic involvement, these components not only address the immediate needs of the community's inhabitants but also support its long-term viability.

Future cities looking to strike a balance between sustainability and growth should use Ecco Ripley's urban development strategy as a model. The project redefines what it means to live in the suburbs by emphasizing economic vibrancy, social inclusion, and environmental care. It proves that extensive growth is possible without sacrificing the landscape's social and ecological integrity. Ecco Ripley is a testament to the ability of careful, integrated planning to create communities that are not only liveable but also resilient and adaptable, while urban areas throughout Australia struggle with the challenges of development.

12.1.1 Planning the New Green: Strategic Foundations and Regional Context

The strategic realignment of growth in Southeast Queensland, one of Australia's fastest-growing regions, is fundamental to the creation of Ecco Ripley. The project, which is situated in the Ripley Valley Priority Development Area (PDA), was designed as a chance to totally rethink the suburban paradigm for a fast-urbanizing area, not merely as a response to population expansion. Fundamentally, Ecco Ripley's design and construction represent a paradigm shift in the way greenfield areas can be developed into people-centered, linked, and sustainable communities that address socioeconomic and ecological demands.

This change is based on a strategic alignment between local goals for sustainable urban futures and development management at the state level. Ripley Valley's status as a PDA allowed for a quicker planning process and the growth of creative planning techniques. The anticipated population expansion in South East Queensland put increasing strain on the region's already-existing urban centers, making the area's development potential both necessary and advantageous. Given the region's projected population growth, especially for younger families, first-time homebuyers, and important workers, a community that would not only provide housing but also improve people's quality of life was required. In this regard, Ecco Ripley became the model for a master-planned community focused on integrated sustainability and long-term liveability.

The goal was to establish a self-contained urban zone that was woven into a broader regional story, not a satellite city. To facilitate decentralized expansion and relieve pressure on inner-city corridors, Ecco Ripley was intended to act as a link between Ipswich and the larger Brisbane metropolitan area. Its function in this polycentric urban layout was to integrate housing, green infrastructure, education, and employment into a logical spatial arrangement that might change over time. This goal necessitated long-term strategic collaboration across sectors and governance levels in addition to design excellence.

Early on, ecological integration became a non-negotiable planning principle. Rather than being viewed as limitations, the site's natural topography, drainage systems, and vegetation corridors were embraced and integrated as strengths. Road networks, housing clusters, and open space systems were oriented according to environmental overlays, and stormwater management and landscape connection were supported by water-sensitive urban design. These initiatives reflected a mental change where the natural environment was viewed as essential to liveability rather than something that needed to be restored after building, and they went beyond simple aesthetic or legal considerations. The end effect is a neighbourhood that literally grows out of the earth, with biodiversity techniques, green spaces, and habitat corridors ingrained in the neighbourhood's fabric.

New demands for urban form were taken into consideration during the planning process, in addition to environmental objectives. It was believed that the conventional low-density suburban sprawl model, which was typified by reliance on cars and inefficient use of space, was unable to address today's issues. Rather, Ecco Ripley adopted a structure that prioritized a variety of housing types, walkability, transit accessibility, and proximity to amenities. Densities are high enough to enable functional public transportation and local services, which is crucial for promoting self-sufficiency and lowering emissions related to commuting, even though they are still relatively modest when compared to metropolitan cores.

Regional policy changes in favor of transit-oriented development were also included in the masterplan. Although the actual expansion of rail services is still a long-term goal, the precinct has been designed with future multimodal connection in mind. Early-stage infrastructure includes bus routes, shared walkways, and bikeways to prevent the neighborhood from becoming overly dependent on cars and to give early residents alternatives to private vehicles. Planning for flexibility, allowing the community to scale, shift, and adapt as infrastructure catches up, was a cornerstone of the strategic vision.

Ecco Ripley's contribution to dispelling suburbia clichés is what sets it apart. The concept incorporates civic nodes, open learning spaces, health precincts, and a town center that is prepared for the future as essential components of urban character, as opposed to depending exclusively on retail anchors or detached housing as structuring devices. They are ingrained from the beginning to foster a feeling of place and belonging that changes with the community; they are not afterthoughts. Understanding how urban trust is developed—slowly and via lived experience—is reflected in the intentional sequencing of development, which begins with lifestyle amenities and green infrastructure before introducing higher-density components.

What makes it unique is Ecco Ripley's contribution to debunking suburban clichés. Instead of relying solely on retail anchors or detached housing as structuring methods, the proposal integrates civic hubs, open learning spaces, health precincts, and a town center that is ready for the future as vital elements of urban character. They are not afterthoughts; rather, they are engrained from the start to promote a sense of place and belonging that evolves with the community. The deliberate sequencing of development, which starts with lifestyle amenities and green infrastructure before introducing higher-density components, reflects an understanding of how urban trust is developed—slowly and via lived experience.

12.2 Ecology, Form, and Function: Building a Sustainable Urban Fabric

The dedication to ecological as a motivating factor for urban form is at the core of Ecco Ripley's master plan. Ecco Ripley creates an urban fabric where sustainability is not an afterthought but a fundamental idea by fusing natural systems with human life, rather than segregating environmental concerns from the logic of building. The physical design techniques of the precinct show how built form, landscape, and infrastructure can all be used to create a community that is effective, flexible, and closely linked to its natural environment.

Ecco Ripley's green spine—a system of parks, walkways, and open areas that function as both social and ecological infrastructure—is one of its most notable features. This green network is more than just a group of reserves dotted around the property. Rather, it functions as an ongoing structure that connects habitat zones, encourages active transportation, controls water flow, and moulds the identity of neighbourhoods. In order to minimize disruption to the land's natural cycles and maximize its ecological functions, the system was planned around the terrain, native vegetation zones, and hydrological patterns that were already in place.

This landscape structure incorporates water-sensitive urban design (WSUD) in a seamless manner. Instead of depending on designed drainage systems that conceal water from the public, Ecco Ripley incorporates permeable surfaces, artificial wetlands, and bio-retention swales into its parklands and streetscapes. In addition to filtering and delaying stormwater runoff, these solutions also foster biodiversity, produce microclimates, and provide the community with visual and educational benefits. In addition to serving a practical purpose, water is aesthetically pleasing, symbolic, and participative, encouraging locals to interact with the natural cycles that support their community.

The form of the precinct is equally centered on walkability and street design. Every residential cluster is connected to important locations, such as parks, schools, and neighbourhood centers, within easy walking distance thanks to a finely crafted network of shared streets, bike lanes, and pedestrian pathways. Streets are made to support life, not just cars. Together, narrower road widths, wide walkways, street trees that provide shade, and traffic-calming techniques create streetscapes that are secure, liveable, and climate-responsive. Through these actions, the suburb moves away from reliance on cars and toward a more harmonious urban ecology of mobility.

Diversity in housing is just another crucial component of Ecco Ripley's long-term plan. The concept incorporates a range of dwelling styles, including townhouses, duplexes, compact lots, and medium-density forms, which are purposefully arranged to create affordability and demographic variety, even if detached homes still make up a portion of the mix. In order to maintain a vibrant and coherent metropolitan community throughout time, this typological diversity promotes social inclusion, economic resilience, and intergenerational living. Crucially, the arrangement of these housing possibilities is influenced by passive design principles, sun orientation, and the goal of maximizing energy efficiency and cross-ventilation. The end effect is a neighbourhood that is environmentally friendly without sacrificing coziness or personality.

Ecco Ripley's reputation as a forward-thinking development is further strengthened by the incorporation of energy-efficient infrastructure and smart utilities. Long-term resilience is established by early investments in fiber-optic communication, subterranean electricity, and effective water and waste management systems. These unseen layers of infrastructure are essential for lowering the ecological imprint of the neighbourhood while simultaneously enhancing daily convenience and standard of living. Invisible and intelligent systems that work well without requiring continual attention or intervention are advantageous to residents.

Beyond its technological advancements, Ecco Ripley is influenced by a place-making concept that is guided by the landscape. The idea takes into account the land's natural contours rather than enforcing a strict urban grid. Public areas are nestled into natural amphitheaters, dwellings open into parks and viewsheds, and street alignments bend to accommodate ridgelines. Beyond just being beautiful, this attention to form strengthens the bonds that people have with their surroundings. It makes it possible for the landscape to perform its natural function of offering grounding, memory, and orientation.

Ecological connectivity is the final thread that binds these strategies together. In Ecco Ripley, the green corridors are ecological continuums that extend throughout the larger Ripley Valley rather than discrete facilities. These corridors connect the site to broader regional ecosystems by acting as carbon sinks, animal passageways, and climate modifiers. The precinct's influence goes well beyond its borders because it was designed with the larger bioregion in mind, making a significant contribution to both climate adaptation and biodiversity conservation.

All things considered, Ecco Ripley's sustainable urban fabric is the outcome of meticulous layering, in which form, function, and environment are interrelated rather than distinct. It serves as an example of how suburban development can be rethought to honour the land, strengthen local communities, and deal with the dual problems of environmental change and urbanization. Ecco Ripley contends that sustainability and liveability are actually the same goal—two facets of a whole urban future—rather than a trade-off.

12.3 Living the Vision: Community Life, Activation, and Everyday Urbanity

Beyond its ecological credentials and infrastructure, Ecco Ripley aims to be a sustainable master-planned community (Mallya 2022). Its identity is rooted in a lived experience that aims to foster not only ecologically conscious behaviours but also strong social ties and a sense of everyday belonging (Cheshmehzangi 2014, 2020). This section examines Ecco Ripley's development as a community-in-formation, where infrastructure, public space, and early placemaking projects combine to build the framework for a thriving and resilient urban lifestyle.

The development has taken a proactive and deliberate approach to community building from the beginning. Ecco Ripley provided essential public infrastructure early in its deployment, in contrast to many greenfield developments that wait for critical mass before adding amenities. Despite being small in its early stages, the Ripley Town Centre acts as a hub for social gatherings and everyday business. In addition to being a place to shop, it serves as a civic forum by holding unofficial gatherings, weekend markets, and neighbourhood festivals that liven up the neighbourhood and promote a feeling of community. The isolation that was frequently experienced in the early years of suburban expansion has been lessened because to this method of establishing a center of gravity right away.

Ecco Ripley's parklands and public areas are likewise intended to welcome life on many levels. The open space hierarchy accommodates a variety of usage patterns, from intimate pocket parks tucked away within residential streets to bigger district parks with recreational fields and play equipment. These parks are designed for interaction, mobility, and collective ownership; they are not just passive green spaces. Bike lanes, nature play spaces, barbecues, and covered seating places promote impromptu interaction and multi-use throughout the day. Crucially, these areas are now the setting for an expanding array of community-led events, ranging from pop-up culinary events and cultural meetings to exercise classes and playdates.

The built form itself reinforces these social possibilities. Homes and streets are built to encourage neighbourly visibility and engagement. There are few front setbacks, which promote porch culture and passive observation. Laneways, common areas, and cul-de-sacs serve as community enclaves that promote micro-neighbourhood dynamics rather than as dead ends. Residential clusters are connected by pathways to public green areas, schools, and stores, facilitating the natural occurrence of social interactions through daily activities.

In addition to its architectural design, Ecco Ripley uses focused social programming to strengthen ties among the community. Early identity and inclusion have been fostered in large part by developer-led programs including welcome days, street barbecues, school partnerships, and art installations. While official institutions like schools and health centers are still being constructed, these programs serve as social glue and soft infrastructure. They show a growing understanding that shared rituals, experienced rhythms, and communal memory—rather than just zoning plans—are what create successful communities.

Despite being staged, civic and educational infrastructure are important elements of this strategy. Long-term dedication to family life and generational continuity is demonstrated by the precinct's establishment of Ripley Central State School and the incorporation of early learning centers into the municipal fabric. In addition to providing services, these institutions are social hubs where parents can meet, children can grow up, and future civic narratives can be shaped. Being close to green areas and pedestrian-friendly streets guarantees that they are a part of everyday life rather than being isolated or on the outskirts.

The adaptability of Ecco Ripley's developing urbanity is arguably one of its most distinctive features. With staging plans, mixed-use potential, and progressive infrastructure deployment intended to change with community requirements, the masterplan purposefully permits flexibility over time. This adaptability shows a sophisticated awareness that no plan can fully account for the complexities of city life. Rather, the precinct encourages a dynamic interaction between form and usage, where areas are planned for both their current use and potential future development as the neighbourhood develops.

Lastly, the everyday routines and common goals of Ecco Ripley's citizens are what write the city's pulse. The choreography of daily life, e.g., children riding bikes to school, neighbours conversing in the park, a local band performing in the town center, is what gives the precinct its spirit, even though the physical surroundings set the scene. In this way, Ecco Ripley demonstrates that the true measure of a sustainable community lies not only in what is built, but in how life unfolds within it.

12.4 Lessons from the Green Horizon: Rethinking Suburban Futures

Ecco Ripley provides an engaging glimpse into Australia's changing suburban urbanization lexicon. Being one of the most ambitious greenfield master-planned communities in the nation, it deviates from traditional suburban models in terms of both form and function as well as its outlook on what suburban living in the twenty-first century might entail. This last portion considers the project's main lessons learned, recognizing its successes while critically examining its limitations. By doing this, it stimulates a more comprehensive discussion about how suburban regions might develop into hubs of social possibility, innovation, and resilience.

Fundamentally, Ecco Ripley questions the idea that suburban growth needs to be low-density, centered on cars, and extractive of the environment. Rather, it shows that master-planned communities may reflect the values of sustainability, walkability, and social cohesion when they are implemented with care and forethought. What a new suburb may look and feel like has been redefined thanks to the project's integrated open space network, emphasis on water-sensitive urban design, and early investment in community activation. In this case, nature serves as a connective tissue rather than a buffer at the boundaries, and density is a structuring logic that permits effective land use, transportation choices, and service provision rather than an afterthought.

The sequencing of Ecco Ripley is among its most noteworthy features. The town center, educational facilities, and community programming were delivered early in the development cycle, creating a framework for lived-in urbanity from the start rather than waiting for population thresholds to justify infrastructure. In greenfield situations, where isolation, delay, and under-servicing frequently result in hollow starts, this is especially important. Ecco Ripley's emphasis on activation rather than abstraction has allowed it to draw in inhabitants, foster trust, and create a thriving community that increases with its physical expansion.

However, there are conflicts in the project's development. Practical constraints will always arise along the blueprint-to-realization process; legislative changes, population shifts, and economic cycles can all have an impact on the delivery timeline and actual results. There are also concerns over the long-term viability of green infrastructure maintenance schedules and the large-scale realization of commercial and mixed-use goals. Furthermore, although while Ecco Ripley has made significant progress in promoting social and environmental sustainability, affordability is still a major problem. Rising land values run the risk of excluding the people that suburban reinvention was originally intended to benefit, as is the case with many new towns in growth corridors.

Larger philosophical issues regarding suburban identity are also brought up by the initiative. Despite being obviously not a typical suburb, Ecco Ripley is a part of the larger cultural and economic factors such as space, privacy, upward mobility, and private ownership, that characterize suburban life in Australia. Therefore, the success of this paradigm depends not just on improving urban form or infrastructure but also on changing perceptions of what suburban life means. Can locals accept smaller lots if they are rewarded with more expansive public areas? Is it possible to view walkability and density as advantages rather than drawbacks? Can master-planned communities develop into really participatory urban societies that transcend beyond private government models?

In this sense, Ecco Ripley is better viewed as a prototype rather than a final product. It encourages education, adaptation, and creativity. Its importance stems from its readiness to challenge long-held beliefs and present a fresh suburban narrative that is in line with Australia's urban difficulties, social ambitions, and climate aims. It implies that suburban futures might represent a variety of shapes, experiences, and connections to place rather of being static or uniform.

The takeaway for planners, designers, and legislators is unmistakable: Australia's suburban future is not ancillary; rather, it is essential. The next phase of sustainable urbanism will be developed not only in crowded inner-city corridors but also in the periphery where expansion is occurring at the fastest rate, as projects like Ecco Ripley serve as a reminder. These regions have the potential to develop into models of inclusive, balanced, and forward-thinking communities if they are directed by a clear goal, extensive consultation, and integrated design.

Ecco Ripley is ultimately a proposition rather than just a location. It poses the question of whether Australia can reinvent its suburban narrative by embracing the past while incorporating modern values, resources, and aspirations. The response will influence a country's urban trajectory in addition to Ipswich's future.

References

Bond S (2010) Best of the best in green design: drivers and barriers to sustainable development in Australia. https://researcharchive.lincoln.ac.nz/entities/publication/281489b0-cc37-44b2-a098-c94f23645e77

Cheshmehzangi A (2014) Spatial syntagma and identity of a place: sensing, relating to, and knowing a place. J Hum Behav Soc Environ 24(7):799–810

Cheshmehzangi A (2020) Reflecting on urban identities. In Identity of cities and city of identities. Springer, Singapore, pp 133–178

Dharmarathna D, Graham P, Bunster V (2025) Unpacking the evolution of green star rating tools and their certification trends in Australian building projects. J Build Eng, 113328

Mallya P (2022) Fine-grained shades of green: integrating permaculture with green infrastructure in master-planned housing communities. University of Technology Sydney (Australia).

Salas K (2017, May) Ipswich is the answer for housing affordability critics. REIQ Journal, 32–33

Sekisui House (n.d.) Ecco Ripley—completed, https://www.sekisuihouse.com.au/communities/ecco-ripley/

Chapter 13
Riverlea Park: Building Community from the Ground up and Designing Scalable Urbanity for Adelaide's Expanding North

13.1 Growing with Purpose: Riverlea Park and Adelaide's Northern Horizon

In one of South Australia's fastest-growing corridors, Riverlea Park is a purposeful reinvention of how urban expansion can bring community, character, and sustainability (Adelaide Plains Council 2023). It is not just another massive residential development. One of the most ambitious master-planned communities in the history of the state is Riverlea Park, which is situated in the northern part of Adelaide in the City of Playford. This project provides a convincing example of how suburban growth may be used as a chance to create futures that are both forward-looking and grounded, as urban forces continue to transform Australia's mid-sized cities.

The goal of Riverlea Park is to create a place where people feel like they belong, not just a place to live. The project covers a wide area that was previously dispersed and mainly underutilized, and it is intended to house approximately 30,000 people through more than 12,000 residences. However, Riverlea Park's thoughtful layering of liveability, sustainability, and infrastructural integration sets it apart from many similar complexes. This is intentional sub-urbanism, moulded by a framework that aims to combine lifestyle with long-term resilience, not suburban sprawl masquerading as planning.

Riverlea Park's location is no accident. Although Adelaide's northern growth corridor has long been recognized as a crucial area for population growth, its development history has frequently been characterized by fragmented projects and communities that rely heavily on cars. The goal of Riverlea Park was to alter this course. While internal planning guarantees the reduction of automobile reliance through walkable neighbourhoods, bike networks, and future-ready transportation choices, its location along important transport connections allow for easy access to Adelaide's central business district. The idea restores the connectedness and spatial efficiency that many suburbs have lost.

A. Cheshmehzangi, *Australia's Urbanity*, Urban Sustainability,
https://doi.org/10.1007/978-981-95-3881-2_13

The masterplan for Riverlea Park is notable for its liberal incorporation of blue and green infrastructure. Over 450 hectares have been set aside for open spaces, which include a central chain of lakes, nature walks, and parklands intended for both recreational purposes and ecological purposes. These elements are necessary to the community's urban logic; they are not only ornamental. In this instance, the landscape anchors and comes before the architectural form. It serves as a psychological and physical link between communities, promoting biodiversity, climate adaption, and well-being all at once.

The precinct's mixed-use character is yet another important asset. Although a considerable amount of housing is being delivered, the development is not exclusively intended for residential use. The proposal incorporates retail strips, commercial clusters, a thriving town center, and health services. A broader understanding that successful places develop not only through structures but also through the networks of daily life is shown in the incorporation of schools, community centers, and sports facilities. In this case, amenity is a place-structuring principle rather than an afterthought.

A mix of conventional detached homes, smaller-footprint homes, and higher-density options that accommodate various life phases and family configurations are all part of the diverse housing type. A more inclusive suburban expansion model, affordability, and social mixing are made possible by this diversity in the housing stock. Riverlea Park incorporates flexibility by eschewing the inclination toward homogeneity, enabling the neighbourhood to develop naturally while maintaining unity and a common identity (Cheshmehzangi 2015).

Importantly, the development shows that master planning in the twenty-first century needs to be more than just land subdivision and aesthetics. Sustainability is an ingrained philosophy rather than a list. Strong environmental buffers, energy-efficient building techniques, and water-sensitive urban design systems are used to guarantee both short-term livability and long-term resilience. These tactics link commercial development interests with public benefit outcomes when combined with well-defined governance systems and phased delivery methods.

Additionally, Riverlea Park signifies a cultural change in Australia's conception of suburban development. Suburban areas were traditionally viewed as either complementing or contradicting the city—dormitory areas where people lived in between commutes. Riverlea Park makes a different suggestion. It demonstrates how the suburbs may serve as hubs for creativity, life, and identity formation (Cheshmehzangi 2020). The precinct seeks to provide an urban pulse commonly associated with inner-city precincts, but in a fresh, greener, and larger setting by carefully balancing openness and density, private residences and public space.

Finally, Riverlea Park is a product of both belief and design. It holds that new communities can emerge with a feeling of unity and kinship. It thinks there might be a significant alignment between suburban growth and sustainability. Additionally, it holds that location is important, both geographically and experientially. Riverlea Park is an example of how future towns can be purposefully designed and how the heart of urban life can emerge even in areas that start out as blank canvases, as Adelaide's northern fringe continues to expand.

13.1.1 Foundations for a Community-Centered Future

Riverlea Park is more than just a house development; it's a modern urban concept with a foundation in the desire to transform South Australian suburbia living. Its cornerstones are a strong vision for long-term liveability, developer foresight, infrastructure coordination, and a thorough alignment of local and state planning policies. In addition to its unparalleled scope, this project is unique in that it prioritizes the development of an economically successful, socially inclusive, and environmentally integrated self-sustaining community from the very beginning.

Riverlea Park's trajectory has been significantly influenced by the policy environment that sustains it. The project is situated in the Playford Growth Area and is governed by strategic urban expansion frameworks (Walker 2024). It is a response to South Australia's overall growth plan, which places a high priority on infill and greenfield development in well-planned, infrastructure-ready corridors. The goal of Riverlea Park was to integrate development into a master-planned framework that provides critical mass and civic purpose, as opposed to permitting fragmented residential subdivisions to flourish in isolation. Together with the state's 30-Year Plan for Greater Adelaide, its strategic location guarantees that the development is a purposeful result of regional vision and spatial governance rather than merely a response to market demand.

As a result, the developer's vision has been crucial in bringing this policy goal to life. The Walker Corporation, which is well-known for its enormous urban projects throughout Australia, came to Riverlea Park with the goal of establishing a generational community rather than just housing. The master plan prioritizes both function and feel by integrating commercial zones, civic institutions, open space systems, and mixed-use precincts into a cohesive whole. Urbanism as an afterthought is not what this is. A story of identity has been woven into the development from the start, emphasizing everyday accessibility, place attachment, and community creation.

Layering has been carefully considered in the planning of land use. Riverlea Park has a hybridized strategy that places schools close to homes, shops within walking distance of households, and green corridors as structural features rather than residual spaces, as opposed to separating functions. This concept encourages a more dynamic rhythm of everyday living where work, leisure, education, and nature are intertwined, in addition to reducing reliance on cars. The reasoning behind the design is obvious: land use is used as a tool to unite people rather than to separate them, and human-scale experience is given priority.

Riverlea Park's excellent integration with infrastructure planning has been essential to its viability and ambition. The synchronization of vital services, such as transportation, utilities, road access, and community facilities, with the precinct's phased roll-out has been made possible by collaborations among developers, local governments, and state authorities. Infrastructure is being incorporated into the very fabric of a housing complex rather than being retrofitted into its periphery. In addition to enabling immediate operation, this coordination prepares the precinct for future economic and population growth.

The development's philosophy has also placed a strong emphasis on economic opportunity. Riverlea Park is intended to be both a residential area and a developing local economy. To keep economic activity local, the precinct incorporates employment land, commercial clusters, and entrepreneurship spaces into the masterplan. at addition to living at Riverlea Park, residents will have the chance to work, produce, and influence the region's changing economy. This promotes a sustainable community paradigm in which social cohesiveness and economic vigour coexist.

Crucially, it is impossible to overestimate Riverlea Park's civic significance. The precinct is planned to provide a civic life that encourages inclusivity, pride, and social connectedness by including facilities for schools, sports centers, libraries, and cultural venues. These amenities are essential to fostering a feeling of community and shared ownership; they are not only ornamental. Riverlea Park avoids the lag that many new communities encounter when social institutions are introduced years after the inhabitants have moved in by integrating civic infrastructure early in the development cycle. Here, bricks and mortar are placed alongside the foundations of community.

Riverlea Park's versatility is what makes it resilient. The development incorporates some flexibility in staging, design typologies, and governance structures in recognition of the fact that cities change in ways that masterplans cannot always foresee. Riverlea Park has incorporated the ability to adapt to changing social norms, technological advancements, and environmental stresses rather than securing the future in a rigid blueprint. Its philosophical and physical underpinnings embrace uncertainty as a natural aspect of the urban environment.

In this sense, Riverlea Park is a prime example of a next-generation suburban planning philosophy, one in which character is not compromised by scale, and community is not sacrificed for expansion. Intention, not just concrete and asphalt, forms its foundation. Its goal is to create a place where urban life flourishes, where the public and private spheres genuinely interact, and where the promise of suburbia is reimagined for the twenty-first century.

13.2 Designing for Everyday Belonging: Streets, Spaces, and the Social Fabric

At the heart of Riverlea Park lies an intentional urban framework designed not only for efficiency or aesthetic appeal, but for cultivating a sense of belonging in everyday life. Street design, public space composition, and housing arrangement are not accidental; rather, they are intentional means of establishing local identity, social cohesion, and spatial clarity. In a development of this size, the little things—like a café on a street corner, a pocket park that encourages people to congregate, or the walkways that link homes to schools—determine whether the area will evolve into a community or merely a collection of homes.

Riverlea Park's street system is designed with permeability, walkability, and readability as top priorities. The master plan provides a hierarchy of streets that blends local intimacy with wider connection, as opposed to relying on recurring grids or looping cul-de-sacs. Main roads are made to support both automobiles and active transportation, with bike lanes and shared pathways allowing locals to go throughout and outside the precinct. Local streets are seen as social spaces; they are smaller, frequently surrounded with trees, and have slower speeds to promote unstructured street life. These decisions allow people to engage with their neighbourhood as a lived experience rather than just a path to follow.

Both structurally and emotionally, public space serves as the precinct's connecting thread. Riverlea Park has strategically utilized its acreage to incorporate civic plazas, linear greenways, and pocket parks into the regular pathways. These areas are not leftovers or residuals; rather, they are prominently located and serve as gathering places for play and informal conversation. Carefully placed seating, shade, water features, and landscaping encourage relaxation and pause. The goal is obvious: design should actively promote community life in addition to accommodating it. By doing this, Riverlea Park creates a backdrop for belonging and turns outside space into an extension of the house.

The built environment's richness is further enhanced by the multiplicity of housing options. By providing a variety of lot sizes, building types, and architectural expressions, the masterplan prevents monotony. There is room for families, individuals, retirees, and first-time homebuyers thanks to the distribution of detached homes, townhouses, and upcoming mid-rise apartments. In addition to providing a market product, this diversity fosters socially diverse and intergenerational societies. This method views housing as a platform for human life in all its diversity rather than as a product to be sold.

The purpose of movement corridors is to unite, not to separate. Shaded walkways, easy access to parks and facilities, and visual cues that promote exploration have all been incorporated into the design to enhance the pedestrian experience. The precinct encourages residents to move through it comfortably and familiarly, whether they are walking to a neighbourhood café, riding their bikes to school, or taking a leisurely stroll to a community center. People can develop a mental map of their location and develop an emotional bond with it by using this ease of mobility, which turns into a type of spatial literacy.

This framework is anchored by neighbourhood focal points. These are not imposing plazas or monuments, but rather commonplace gathering places where people naturally congregate, like a neighbourhood grocer close to a daycare facility or a green square surrounded by houses and seats. Because of their thoughtful placement at walkable distances, these hubs support a localized rhythm of engagement. They also act as holding spaces for community infrastructure, such as gyms and meeting rooms, which promote regular involvement and civic engagement.

Riverlea Park's design supports life in all its complexity rather than trying to dictate how it develops. It is believed that the areas between buildings are equally as significant as the actual buildings. Every aspect contributes to the texture of street-level living, including the positioning of doorways, the transparency of façades, and the presence of porches and balconies. These modest yet impactful actions strengthen feelings of security, friendliness, and concern. Ultimately, the physical layout of Riverlea Park is intended to support a way of life rather than to make a statement. It acknowledges that a sense of belonging is gradually developed via recurring interactions, everyday routines, and common areas that entice individuals to stay, connect, and come back. The precinct thus provides a model for suburban design that goes beyond efficiency and beauty to a more profound comprehension of how urban form may fulfill the human desire for place and connection.

13.3 Urban Life in Practice: Activation, Identity, and Local Pulse

In addition to its master plan or constructed form, Riverlea Park's goal of creating a thriving, community-driven precinct is rooted in its lived rhythms, which are the patterns of activity, interaction, and shared meaning that give the development its vitality. The project has understood from the beginning that successful urban life cannot wait until the last brick is placed. Rather, it must start with a deliberate dedication to community involvement, early activation, and the establishment of local identity. These initiatives provide the framework for a neighbourhood that strives to be more than just liveable—it wants to have purpose.

The intentional placement of early hubs is one of the main tactics supporting the growth of life at Riverlea Park. These hubs, which are frequently supported by businesses or public services like community centers, daycare centers, and cafes, offer more than just useful amenities; they give locals a cause to get together, socialize, and share space. The initiative builds a basis for informal connection and neighbourliness and lessens the sense of isolation that is typical of new estates by making sure that these hubs are operational even in the early stages of construction.

This local vibrancy is further supported by investments in the public space. Open green spaces, natural corridors, plazas, and walking paths are viewed as dynamic social infrastructure rather than as decorative afterthoughts. From families with kids to retirees looking for regular social interaction, they welcome all demographics with their pop-up events, outdoor fitness classes, weekend markets, and seasonal programming. In order to ensure that people of all abilities may engage in daily activities and contribute to the precinct's social ecosystem, these areas were also created with accessibility in mind.

Integrating educational institutions is essential to establishing a sense of community. Plans for schools inside or close to the precinct aim to integrate educational institutions into the very fabric of the neighbourhood, not only to accommodate future population development. Schools act as hubs for intergenerational communication, family activities, and the development of a sense of community. Additionally, their integration creates a continuous flow of civic events that connect individuals to their locality and enhances walkable catchments.

Riverlea Park's emphasis on inclusive placemaking and cultural programs is equally important. To make sure that activation techniques represent a range of customs, values, and goals, the precinct's developers and planning teams have collaborated with local governments and community organizations in recognition of the increasing diversity of its resident base. Small but impactful manifestations of this dedication to representation include multilingual signage, art workshops, cultural festivals, and community storytelling programs. In addition to lessening alienation, these initiatives actively promote a sense of collective ownership in the development of the precinct's narrative.

Riverlea Park's identity grows along with it. This project permits its identity to be co-authored over time by those who live, work, and congregate there, in contrast to developments that aim to impose a fixed brand or character. Early residents are actively creating the intangible but crucial layer of place culture by their involvement in events, their usage of public areas, and their unofficial customs—like meeting at a favourite picnic area or taking the same morning stroll. This kind of placemaking recognizes that customs, tales, and the gradual accumulation of common experience are just as important in creating a sense of community as physical structures.

This natural development of place identity is further strengthened by local collaborations. Riverlea Park's planners have made it possible for the precinct to experience a sense of belonging by working with small businesses, local artists, civic organizations, and non-profit service providers. This relational approach to development aids in bridging the divide between bottom-up inhabitation and top-down planning. It guarantees that the precinct stays linked to the northern growth corridor of Adelaide rather than developing independently.

Perhaps most importantly, Riverlea Park's approach to urban life values continuity over spectacle. It relies on fostering everyday interaction, impromptu get-togethers, and consistent engagement with the local environment rather than relying on one-time events or advertising activities (Gurran et al. 2021). By doing this, it creates a living environment as well as a liveable one, where the rhythms of daily life are supported by welcoming spaces, inclusive programs, and tactics that address the complicated realities of urban communities in the twenty-first century. In this sense, Riverlea Park is fostering communal life from the ground up rather than only constructing residences. It recognizes that a precinct's success depends on how it is populated, modified, and cherished over time, in addition to how it is planned.

13.4 Lessons in Scalable Urbanism: Sustainability, Equity, and the Long-Term View

An important test case in Australia's changing urban landscape is Riverlea Park, which aims to combine suburban growth with environmental responsibility, urban sensitivity, and community inclusivity. Being one of the biggest master-planned communities in South Australia's history, it provides more than just a place to live; it is a declaration of potential that massive expansion may still be based on long-term social value, ecological awareness, and human needs.

One of the clearest lessons from Riverlea Park is the importance of scale-conscious planning. The project's enormous scope, which includes thousands of future residences, town centers, parks, schools, and work areas, necessitates a prompt and flexible infrastructure delivery schedule. In this case, infrastructure staging turns into a key component of sustainable urbanization rather than only a technical issue. A sense of functioning and momentum has been shaped in large part by the early implementation of community centers, water-sensitive urban systems, and integrated transportation links. Communities are more likely to develop with stability and trust when amenities are introduced before or alongside residents, rather than years after. In this manner, urban life is fostered through early investment in the social and ecological backbone of the development, while scale is controlled through sequencing.

However, growth also adds complexity to governance. Coordination between many stakeholders, including developers, local governments, service providers, and potential residents, who all have an interest in determining the character and purpose of the community, is one of the new issues facing precincts like Riverlea Park. Community consultation must continue to be a continuous process rather than a one-time occurrence, and the governance of design purpose must be safeguarded without becoming inflexible (Ross 2018). The lessons learned here indicate that community trusts, covenants, and design rules can be effective instruments for guaranteeing a constant standard of constructed form and public space—while simultaneously allowing for the growth of local culture and grassroots innovation.

Another important factor is equity. Riverlea Park's inclusion of a variety of dwelling styles, including townhouses, detached homes, and possible medium-density choices, sets a precedent for socially mixed communities at a time when housing costs are rising across metropolitan areas. However, affordability involves more than just price points. Ensuring that every person, irrespective of their origin or income, has access to excellent public space, transit connectivity, educational opportunities, and civic infrastructure is the foundation of true equity. Additionally, an equitable paradigm that can withstand the trend toward spatial inequality or monocultural enclaves must be scalable.

Riverlea Park also suggests what a next-generation suburban form might entail from a sustainability standpoint. A noticeable change in the behaviour of greenfield development is provided by its dedication to maintaining biological corridors, utilizing sustainable drainage systems, and designing around natural features rather than removing them. The project aims to compose with the terrain rather than overwrite it. For other rapidly expanding areas, this strategy offers an important lesson: environmental logic should be the cornerstone of urban logic rather than its secondary.

The way Riverlea Park views the suburban fringe as a place of reinvention and frames suburbia as more than the legacy of the past may ultimately be its greatest gift. The concept breaks down the old dichotomies of urban and suburban, dense and scattered, natural and constructed, by incorporating walkability, community development, local economics, and environmental preservation into a comprehensive master plan. It offers a convincing case for the scalability of good urbanism—not by weakening its tenets, but by firmly establishing them in local context, thoughtful design, and a long-term vision of what a sustainable community life ought to be.

References

Adelaide Plains Council (2023) Growth and background Paper, https://www.apc.sa.gov.au/__data/assets/pdf_file/0032/1345199/Growth-Background-Paper-May-2023.pdf

Cheshmehzangi A (2015) Urban identity as a global phenomenon: hybridity and contextualization of urban identities in the social environment. J Hum Behav Soc Environ 25(5):391–406

Cheshmehzangi A (2020) Reflecting on urban identities. In: identity of cities and city of identities. Springer, Singapore, pp 133–178

Government of South Australia (2020) Application on notification—Impact assessed development, document by the department for trade and investment, main report available from https://plan.sa.gov.au/__data/assets/pdf_file/0007/1233736/Buckland-Park-Riverlea-Township-AEIS-Part-1-Main-Report.pdf

Gurran N, Forsyth A, Darcy M, Searle G, Buckle C (2021) Population growth, regional connectivity, and city planning—international lessons for Australian practice. AHURI final report

Ross S (2018) Strategies for more inclusive municipal participatory governance and implementing Un-Habitat's new urban agenda: improving consultation and participation in urban planning decision-making processes through rapid ethnographic assessment procedures. Can. B. Rev. 96:294

Walker (2024) Response document to the Riverlea major development, 80 pages document https://plan.sa.gov.au/__data/assets/pdf_file/0008/1509146/Buckland-Park-Riverlea-Township-Response-Document.pdf

Chapter 14
Tracks to Community: The Urban Promise of Hills Showground Station Precinct in Sydney's Northwest

14.1 Stations as Starters: Hills Showground and the Rise of Transit Urbanism

In Sydney's suburban development, the conversion of Hills Showground into a bustling, mixed-use zone represents a turning point. Once representative of low-density, automobile-focused suburbia, Castle Hill is now the focal point of one of Australia's most ambitious transit-oriented development (TOD) initiatives. Its rise is a purposeful act of urban repositioning rather than just a result of infrastructure growth; the station is transformed from a mobility hub to a driving force for density, diversity, and a new cadence of urban life.

A showpiece project of the larger Northwest Metro plan, the Hills Showground Station Precinct aims to change Sydney's growth trajectory by focusing development along new transit links (NSW Government, n.d.). This initiative makes use of the station's position to start a finely-grained urban environment, in contrast to conventional suburban railway expansions that depended on park-and-ride methods. Here, a lifestyle that emphasizes walkability, liveability, and a harmonious interaction between constructed form and public space is supported by easy access to high-frequency transit.

The station is only a short stroll from the district, which has been thoughtfully designed to include a mix of residences, shops, green areas, and community services. Hills Showground stands out for its dedication to planning from the outside in, as seen by pedestrians. The project seeks to create a cohesive community framework, where laneways, plazas, and busy ground-floor frontages impact the everyday experience of commuters and residents alike, instead of haphazardly piling density. Streets serve as locations of continuity and encounter in addition to being connectors.

The precinct's approach to the problems of urban sustainability and mobility is fundamental to its identity. Hills Showground aims to normalize car-free lifestyles in contrast to Castle Hill's long-standing reliance on private vehicles. This concept is supported by important components including integrated bike networks, direct access

A. Cheshmehzangi, *Australia's Urbanity*, Urban Sustainability,
https://doi.org/10.1007/978-981-95-3881-2_14

to metro services, restricted parking ratios, and a focus on last-mile connectivity. A more developed stage of suburban development thinking is shown by this change in mobility logic from dependence to choice, when public transportation becomes the cornerstone of place-making rather than an afterthought.

The station itself serves as an urban anchor, being both central and welcoming without being intrusive. It serves as the spatial and symbolic hub of the developing precinct and has a design language that prioritizes openness, public engagement, and ease of navigation. The ideas of height gradation, sun access, mixed tenure, and activated margins are guiding the construction of a number of residential and commercial developments surrounding it. Even in a relatively high-density footprint, human-scale transition zones are made possible by this development choreography.

However, the change extends beyond transportation infrastructure and bricks. It reflects a broader goal: to support the development of communities alongside actual construction. Early activation techniques, such as temporary parks, community input, public art experiments, and pop-up events, were incorporated into the Hills Showground planning process with the goal of fostering emotional investment from potential residents and surrounding neighbours (Ethos Urban 2025). The development approach here recognizes the importance of front-loading community identity rather than waiting for population thresholds to justify services and amenities.

By doing this, the precinct also provides commentary on how the concept of suburban attractiveness is changing. Nowadays, accessibility, adaptability, and cultural vibrancy are more important than just quiet neighbourhoods and separate homes. With its green infrastructure and smooth mobility, Hills Showground presents itself as a location that combines social interaction, affordable living, and excellent design. It becomes a new urban promise within reach for professionals, downsizers, and younger families: more expanded choice, less sacrifice.

This project's demonstrated effect has broader implications. Hills Showground offers a replicable model as other areas of Sydney—and Australia—face challenges related to infill, densification, and infrastructure coordination (MacDonald 2016). It demonstrates how metro investments can be used to support comprehensive city planning as well as transportation outcomes. Additionally, it emphasizes how stations have the power to alter entire growth trajectories when viewed as the start of urban talks rather than their conclusion.

In the end, the Hills Showground Station Precinct questions established divisions between liveability and transit, urban and suburban, and infrastructure and identity. It makes the claim that, when location, planning, and people are in harmony, the suburban fabric of the future will not be one of low expectations and gradual change, but rather one in which urbanity can thrive in new geographies. Hills Showground is a clear, palpable, and developing example of what next-generation suburban urbanization can look like as Sydney's northwest corridor continues to urbanize.

14.1.1 From Paddocks to Platform: Planning a Connected Urban Future

Hills Showground's transformation from a low-density, outlying area into a linked, future-focused urban center is the result of a concerted planning vision, consistent infrastructural investment, and a growing political resolve to rethink the outer suburbs of Sydney's role. What was previously a vast area of parking lots, showgrounds, and semi-rural paddocks has been transformed into a model for suburban urbanism. It is organized around the promise of community-centered expansion, transit accessibility, and diverse housing rather than cul-de-sacs and road hierarchy.

The deliberate choice to expand Sydney's Metro line into the Northwest Growth Area, an infrastructure commitment that recalculated the growth logic for the entire region (NSW Government: Planning, n.d.), is at the core of this change. Hills Showground Station was picked for a reason: it was ideally situated to show how mobility infrastructure may open urban opportunities because it straddles the urban–rural divide. The station turned into a purposeful anchor for a new type of place-making, where walkability influences everyday life, density is managed precisely rather than compromising, and mobility determines land value.

A major factor in this change was the government's vision. The Hills District became a crucial growth corridor in the city's decentralization goal after Sydney's Metropolitan Strategy was released, which placed a strong emphasis on "30-min cities." The precinct was designated not just for home supply but also for the development of a full neighbourhood, providing employment, services, education, and green infrastructure in a catchment area that is walkable. In order to do this, state agencies, transportation authorities, and local councils had to abandon compartmentalized planning methods and adopt a cooperative paradigm that aligned zoning, infrastructure provision, and design quality standards from the beginning.

Repositioning the precinct began with the rezoning of the property around the station. With mixed use overlays, height flexibility close to the station, and a number of urban design constraints meant to strike a balance between expansion and liveability, what had long been trapped in low-density patterns—with big lots, disconnected roads, and single-use zoning—was suddenly redesigned. In order to guarantee ground-floor vibrancy and pedestrian scale, this involved requiring street-level permeability, lowering automobile parking ratios, and requiring active frontages.

Importantly, community consultation procedures that aimed to reconcile expectations from the past and the future also influenced the development's course. Castle Hill and the surrounding Hills region's residents had deep relationships to their homes, which were frequently based on the suburban values of seclusion, greenery, and detached homes. By placing a strong emphasis on the preservation of current parklands, the establishment of new open spaces, and the early development of amenity infrastructure during the build-out phase, the planning process attempted to reconcile these ideals with newly emerging urban objectives. Planners presented densification as evolution, a move toward a more cohesive, service-rich form of communal life—instead of erasure.

In order to increase operational efficacy and public trust, infrastructure sequencing was essential. Authorities made sure that public transportation access was not an afterthought but rather a structural feature from the beginning by prioritizing the station's construction alongside the early phases of residential and municipal development. The development of a walkable urban core, where businesses, educational institutions, and services could flourish without having to wait for full population thresholds, was also aided by this strategy. Even in its early years, the precinct was passable and liveable thanks to the integration of bike and pedestrian networks, shaded public areas, and connecting lanes, which created a clear framework for gradual development.

The investment in social infrastructure was equally significant. By requiring developers to help build community centers, educational institutions, and recreational areas, planning frameworks incorporated daily living into the basic logic of development. These were presented as essential components of the neighbourhood's character, not as long-term pledges or optional frills. Hills Showground did this by changing the discourse surrounding suburban development from one of land speculation to one of purposeful, human-scaled place-making.

It is impossible to overestimate the importance of landscape in this planning goal. Early master planning included urban greenery, stormwater-sensitive design, and natural corridors to offer civic identity and ecological integrity. These components, which frame the experience of streets and buildings, direct pedestrian pathways, and provide areas for relaxation, play, and community gathering, were viewed as essential components of the precinct's urbanism rather than as obstacles or buffers.

A precinct that reinterprets what suburban intensification can entail in Australia is the result of this meticulously planned and policy-driven endeavour. Hills Showground serves as an example of how well-considered planning, integrated infrastructure, and cooperative governance can transform sleepy margins into vibrant hubs where the station serves as a starting point rather than a boundary. By doing this, the precinct presents a fresh idea of development, one that is more thoughtful, more interconnected, and ultimately more civic.

14.2 Vertical Village Thinking: Built Form, Public Space, and Density Design

A new era of suburban development is typified by the Hills Showground Station Precinct, where vertical density is intended to facilitate rather than to dominate. The concept of a "vertical village", a high-density, mixed-use setting that seems scaled to people, not simply property values; it is at the heart of the precinct's built form design (Rashid 2011). Hills Showground is a well-balanced mix of residential, commercial, and civic facilities, intended to promote community life and spatial clarity within a small footprint. It is very different from the solitary apartment buildings or impersonal corporate parks of previous suburban attempts.

The precinct's architectural style strikes a balance between modulation and massing. The height of the buildings varies according to how close they are to the station core and taper off toward the outskirts of the surrounding neighbourhoods. The precinct uses a contextual approach to the built form rather than emphasizing a single iconic structure. This involves spreading mass through podium-tower combinations that permit light access, permeability, and street-level human-scale engagement. While the top levels house a variety of housing types, including apartments that cater to varied demographics like downsizers, young professionals, and growing families, the ground floors are lively with shops, cafes, and community services.

The prioritization of public space is essential to this vertical strategy. Buildings are connected and easily navigable through the use of green corridors and pedestrian paths. The network of open spaces is not accidental; rather, it is purposefully created to fulfill many functions including areas for play, mobility, relaxation, and casual socializing. To provide regular opportunities for green space and civic stop, the precinct is filled with shared laneways, pocket parks, and little plazas. The precinct's dedication to ecological integration and liveability is further reinforced by the planning logic that guarantees both locals and tourists can always find a green refuge within a short walking distance.

The inclusiveness principle has also served as a guide for public realm design. A friendly, accessible, and secure pedestrian environment is influenced by a variety of factors, including lighting design, furniture selection, pavement widths, and shade availability. Instead of having monolithic façades or blank walls, street-level interfaces are made with articulation, transparency, and usage that flow into the public area. The use of street trees, bioswales, and water-sensitive design features enhances the aesthetic appeal and environmental sustainability of hard infrastructure by creating a landscape with layered experiences and soft edges.

The precinct's gravitational center is the transportation hub itself, which is anchored by Hills Showground Station. However, it has a larger symbolic purpose than just being utilitarian; it represents a new urban rhythm. The station precinct is intended to be a destination in and of itself, not just a place to arrive or exit. It is a plaza surrounded by shops and public facilities, bustling with commuter traffic, and layered with chances for rest, conversation, and engagement. The precinct's fundamental character as a paradigm of transit-oriented development is strengthened by the smooth intermodality between the metro, buses, walking, and bicycling.

Tight integration of mixed land uses has been implemented to provide balanced and ongoing activity. The careful placement of commercial spaces ensures that housing is within walking distance of employment opportunities. Community and educational facilities, such as planned daycare centers, co-working spaces, and health services, are intended to serve as the neighbourhood's focal point. The precinct creates an ecology of everyday life where people live, work, and play in close proximity to one another rather than dividing functions across zones.

Crucially, readability is not sacrificed for the architectural form's verticality. Wayfinding has been embedded into the design language through consistent materials, sightlines, signage, and thresholds. Streets are urban spaces with distinct personalities, rhythms, and public uses rather than highways for automobile traffic. Residents will see the precinct as a pedestrian network of connected locations rather than as a collection of intimidating structures thanks to the built form, which frames rather than overwhelms.

Design and density don't have to be mutually exclusive, as the Hills Showground Station Precinct shows. Vertical urbanism can be reimagined for Australia's suburban futures with deliberate use layering, precise scale calibration, and careful edge articulation. It is a human-faced paradigm of compact expansion, where density enhances rather than lessens the rhythms of urban life and design transforms aspirations into everyday reality.

14.3 Living at the Node: Placemaking, Activation, and Urban Mobility

The lived experience is essential to the Hills Showground Station Precinct's transit-oriented development model's success. Mobility is closely linked to social presence, placemaking, and daily interaction in this setting, which is not just meant for movement but for life in motion. Hills Showground, a crucial stop on the Northwest Metro line, has been designed to provide connection—between people, events, and space—rather than just access.

With its integrated network of pathways, shared bike lanes, and short walking distances between residential buildings, public spaces, and the metro station, the precinct's layout promotes active lives. The precinct reverses the paradigm to prioritize public transportation and active mobility rather than promoting reliance on private vehicles, which is a defining feature of conventional suburban development. This change gives locals a sense of possibilities and closeness while promoting a slower, more grounded pace of daily life in an otherwise rapidly expanding area.

In Hills Showground, placemaking includes more than just building infrastructure. It is ingrained in the deliberate design of settings to encourage use and produce meaning. The layout incorporates green forecourts, private lanes, and public plazas to provide areas for people to live, congregate, and travel around at their own pace. These areas have two purposes: they improve permeability throughout the precinct and offer recreational and social opportunities. Even in a high-density context, the precinct fosters a sense of familiarity by designing readable, cozy spaces that seamlessly move between various urban intensities.

Having mixed-use ground-floor programming is a major factor in creating urban vitality. The lower levels of buildings are activated by retail strips, cafes, co-working spaces, and community services, which provide a variety of reasons for people to interact with the area throughout the day. Crucially, these are viewed as civic attractions that contribute to the local identity rather than as merely commercial extras. Social encounters and casual meetings are shaped by the natural flows of people who congregate around important movement corridors and plaza areas. Here, daily activities take place in ways that make staying as much of an option as commuting.

The temporal diversity of urban life is also reflected in the precinct (Cheshmehzangi 2016). The area stays welcoming after the regular workday thanks to carefully considered lighting plans, nighttime security measures, and evening activation. Recognizing that the precinct will need to change over time to accommodate its expanding community, the area is kept flexible and socially elastic through event programming and temporary installations. These usage rhythms are essential to creating a living precinct that provides a beat, a pulse, and a tempo of shared experience in addition to architectural presence.

Hills Showground stands out due to the way its mobility systems and physical infrastructure are integrated with the social potential design. The station itself is portrayed as an anchor for urban experience rather than a solitary point of arrival and departure. Place is enhanced rather than diminished by movement because transit is integrated into the neighbourhood's visual, physical, and social fabric. Going somewhere is the same as being somewhere significant.

Additionally, the experience of accessibility and inclusion is reinforced when dwellings are close to amenities, stores, green areas, and transportation. The precinct offers a variety of apartment styles, pricing points, and lifestyle amenities to accommodate intergenerational living. Building a socially vibrant and resilient community that reflects the evolving suburban landscape of Greater Sydney requires this diversity. Economic opportunities are also expanded by walkable transportation, which frees locals from the burden of relying on cars to reach services and employment marketplaces throughout the metropolitan area.

Hills Showground Station Precinct demonstrates how transport-oriented development may go well beyond function by focusing density on transit and incorporating liveability into design choices. It develops a lifestyle offering based on shared identity, sustainability, and closeness. In this case, the transportation node turns into a hub for urban interchange, where the movement of people and ideas creates a more vibrant, inclusive urban environment.

14.4 Beyond Infrastructure: Lessons for Integrated Suburban Futures

The Hills Showground Station Precinct provides an engaging example of how purposefully built infrastructure may revolutionize suburban development. This concept puts infrastructure at the center of an integrated urban experience rather than considering transit as an afterthought or a stand-alone mobility instrument. Lessons can be learned not only from its constructed results but also from a more comprehensive strategy for reimagining what suburbs can be when liveability, density, and transit are all coordinated.

The importance of early coordination between governmental levels, transportation agencies, urban planners, and private developers is among the most noteworthy lessons learned from the precinct. The common goal that guided development from the beginning—one that acknowledged the significance of integrating transportation infrastructure into a larger urban framework—is what makes Hills Showground so successful. In suburban development, where piecemeal growth can result in fragmentation and inefficiency, this degree of strategic congruence is frequently lacking. Hills Showground, on the other hand, is a prime example of the importance of integrated long-term planning that integrates social infrastructure, public realm design, mobility, and land use into a single coherent plan.

The value of adaptability and flexibility serves as a second lesson. Hills Showground was planned with future development in mind, even though the built environment provides a certain amount of permanency (The Hills Shire Council 2018). Staged development, scalable density, and flexible space usage are all made possible by the precinct. Because public spaces are not unduly restrictive, they can be used for new purposes as community needs change. This strategy recognizes that suburban areas are dynamic and that long-term resilience depends on their capacity to adapt to changes in the population, economy, and climate.

The strategy for housing choice and diversity offers yet another insight. Hills Showground seeks to provide a diverse residential mix rather than homogeneous apartment buildings or a single typology, allowing a wider range of people to dwell close to amenities, services, and transit. The development creates an urban ecology that is socially inclusive by bringing a variety of age groups, household kinds, and income levels into the precinct. By doing this, it contributes to redefining the suburban narrative by providing a low-density, car-dominated option to living in a neighbourhood while preserving aspects of accessibility and neighbourhood familiarity.

Crucially, transportation plays a fundamental role in Hills Showground rather than a symbolic one (NSW Legislation 2019). The metro station is reinvented as a civic anchor that structures the precinct's existence rather than just serving as a gateway to other places. In suburban settings, where accessible access to dependable, high-frequency transportation opens up new opportunities for mobility, work, and lifestyle, this strategy reinterprets what proximity means. Additionally, it repositions the suburb as a local node contained in a larger metropolitan network, moving it from a location of separation to one of convergence.

Lastly, Hills Showground Station Precinct shows us that suburban development involves more than simply infrastructure and density addition. It is about designing with human existence in mind, establishing communal conditions, and integrating purpose into planning. The lessons learned from this district remain relevant and instructive as Australian towns continue to struggle with environmental demands, growth challenges, and changing societal expectations. They serve as a reminder that future suburbs ought to be more cohesive, liveable, and representative of the intricate rhythms of daily urban life.

References

Cheshmehzangi A (2016) Temporary and temporality: public realm regeneration through temporary events. J Urban Regen & Renew 10(1):58–72

Ethos Urban (2025) Planning proposal: proposal to amend the hills local environmental plan 2019, https://www.google.com/url?sa=t&source=web&rct=j&opi=89978449&url=https://hillshaveyoursay.thehills.nsw.gov.au/101630/widgets/466593/documents/311548&ved=2ahUKEwiws_n04MWOAxXD1jQHHbuPGS4QFnoECA4QAQ&usg=AOvVaw1piR-cVtLdTBtEJJPB51YJ

MacDonald H (2016) "Green, global, and connected": can Sydney solve its metropolitan governance problems?. In: transforming distressed global communities. Routledge, pp 211–230

NSW Government (n.d.) Showground station: Sydney Metro Northwest Urban Renewal Corridor, https://www.planning.nsw.gov.au/plans-for-your-area/priority-growth-areas-and-precincts/sydney-metro-northwest-urban-renewal-corridor/showground-station

NSW Government: Planning (n.d.) Sydney Metro Northwest Urban renewal corridor: priority growth areas and precincts, https://www.planning.nsw.gov.au/plans-for-your-area/priority-growth-areas-and-precincts/sydney-metro-northwest-urban-renewal-corridor

NSW Legislation (2019) The hills local environmental plan 2019, https://legislation.nsw.gov.au/view/whole/html/inforce/current/epi-2019-0596

Rashid MM (2011) Liveability and sustainable design in higher density inner urban housing in Sydney. Doctoral dissertation, UNSW Sydney

The Hills Shire Council (2018) Public domain plan—showground station planned precinct, https://www.thehills.nsw.gov.au/files/sharedassets/public/ecm-website-documents/page-documents/building/dcp/public_domain_plan_-_showground_precinct_-_post_exhibition_fp223_68_pages.pdf

Chapter 15
Yarrabilba and the Pulse of Sustainable Suburban Growth: Designing the Sustainable Satellite City in Logan, Queensland

15.1 A City of Intention: Yarrabilba and the Vision of Balanced Growth

Yarrabilba is a city of purpose rather than merely a master-planned community. Yarrabilba, one of Queensland's most ambitious greenfield urban developments, represents a significant change in Australia's approach to rethinking suburban growth for the twenty-first century. Situated in the rapidly expanding Logan corridor, which connects Brisbane and the Gold Coast, the precinct holds a unique position as a testbed for sustainable urbanization and a recipient of regional growth (Alam et al. 2014). It is intended to serve as a proactive example of how to construct a complete urban center around the principles of liveability, environmental integration, and civic inclusion rather than just as a reaction to demographic pressure.

Yarrabilba is unique because of the scope and consistency of its goals. It is anticipated that the development will house about 50,000 people in over 20,000 residences with a wide range of auxiliary uses, including community centers, parks, schools, local business clusters, and health facilities (Logan City Council, n.d.). However, this vision is really qualitative rather than merely quantitative. Within a framework that seeks to strike a balance between environmental preservation, economic opportunity, and human well-being, the project places a high priority on green infrastructure, walkability, social services, and local employment.

Most importantly, Yarrabilba is situated within a larger regional perspective. The challenge for Southeast Queensland is to accommodate its rapidly expanding population without repeating the pitfalls of disjointed, car-dependent sub-urbanism. Yarrabilba is a purposeful turn in this regard. With the help of robust transit connections and regional planning that aims to concentrate expansion into well-served nodes, it is intended to develop into an independent urban center. Because of its site, which was formerly a combination of farming and forest, it has also necessitated a careful approach to ecological preservation, making integrated stormwater systems, green corridors, and wildlife buffers essential components of its urban design.

A. Cheshmehzangi, *Australia's Urbanity*, Urban Sustainability,
https://doi.org/10.1007/978-981-95-3881-2_15

However, Yarrabilba is ultimately about creating a way of life, which goes beyond its spatial logic and planning frameworks. The precinct has prioritized community development and place-making as equal partners to infrastructure and growth since its founding (Delfin Lend Lease 2004). Its delivery has been characterized by early activation, with cultural programs, civic centers, athletic fields, and schools all aimed at promoting identity and connection long before the population curve peaks. Enabling people to live meaningfully—in an environment where every day routines, social interactions, and environmental surroundings feel unified, respectable, and forward-looking—is the aim, not just housing them.

The built form also exhibits this change. The precinct is constructed with separate neighbourhoods, each with access to local natural space, public realm interventions, and a variety of dwelling typologies, as opposed to monolithic sprawl. Density is measured as a lived characteristic that accommodates a variety of tenures and lifestyles, walking distances, and shared spaces rather than as an abstract quantity. For instance, the developing town center aims to be a genuine gathering place rather than a transactional retail strip by combining civic presence with commercial vibrancy.

It is noteworthy that Yarrabilba has also relied on collaborations that go beyond the traditional developer-government dynamic. The precinct's DNA has included partnerships with social entrepreneurs, educational institutions, healthcare organizations, and neighbourhood associations. This spirit of cooperation recognizes that cities are fostered rather than merely constructed. Additionally, building a new urban center from the bottom up necessitates paying attention to connections, stories, and everyday experiences in addition to roads and bricks.

Yarrabilba provides a window into the future of urban growth in Australia's regional corridors in this way. There are still concerns over affordability, economic self-sufficiency, and long-term transport integration, thus it is not without hazards and criticisms. Its resistance to accept that inevitable fate, however, is what gives it its true significance. Rather, it takes back the suburban frontier as a place for community creation, experimentation, and design. It demonstrates that expansion may be an opportunity rather than a burden if it is approached with consideration and foresight.

We are reminded that planning is more than just technical work; it is cultural effort when Yarrabilba is framed as a city of intention. It involves expressing the kinds of communities we hope to inhabit and the futures we are prepared to create collectively. Yarrabilba is a long-term project that is still in its infancy. However, it is already making a strong statement: the next wave of Australian urbanism might emerge not in our historic districts but rather at the thoughtfully designed boundaries where location meets design and expansion meets intention.

15.1.1 *Growth Meets Ecology: Strategic Planning and Environmental Mandates*

The creation of Yarrabilba as a sprawling master-planned community was never solely a reaction to the need for housing; rather, it was an intentional effort to balance the need for ecological care with the urgency of urban growth. The foundation of the strategic planning framework that supported the project from the start was this harmony between ecology and growth. Situated in Logan City, Yarrabilba is situated in the nexus between Southeast Queensland's population demands and the requirement to preserve delicate environmental corridors. In order to create a new form of suburban development in Australia, its creation represents a conscious connection of legislative ambition, environmental objectives, and infrastructural investment.

Under Queensland's planning regime, Yarrabilba was named a Priority Development Area (PDA) at the policy level, providing the chance to expedite delivery while still adhering to state-level sustainability and liveability goals. Although there was a lot of planning scrutiny associated with this status, there was also more freedom to construct outside of established zoning guidelines. Instead of reactive patchwork growth, it allowed the development to be organized around long-term land use plans, infrastructure sequencing, and environmental thresholds (Urban Land Development Authority 2010). Land availability and closeness to established metropolitan centers were important factors in the site's strategic selection, but so was the opportunity to start from scratch and create a community where ecological principles might influence the shape and pace of development.

Environmental regulations had a significant impact. A large portion of the land was once a mix of cleared farming and wooded regions; this is biodiversity-rich land that is characterized by native flora, riparian corridors, and watercourses. During the masterplanning process, these natural systems were seen as guiding principles rather than limitations. With the help of a green infrastructure framework, key natural elements like the Logan River watershed and native habitat zones were preserved and incorporated into the urban landscape (Bajracharya and Hastings 2020). Yarrabilba was planned around nature rather than over it, utilizing its natural resources as ecological, recreational, and social infrastructure. Water-sensitive urban design is supported by systems of retention basins and bioswales, linear parks, wildlife corridors, and flood-sensitive land designated for passive uses.

The necessity of constructing almost 20,000 residences, educational facilities, healthcare facilities, shopping malls, and employment districts had to be weighed against these environmental concerns (Queensland Government 2025). The development is being implemented in phases, with infrastructure and public realm investments coming after each wave of residential growth. In this case, the plan was based on phasing. This makes it possible to calibrate service capacity, continuously evaluate environmental performance, and modify the urban design to meet changing needs. In addition to providing services to the city's immediate people, infrastructure investment—particularly in the areas of water, transportation, energy, and digital connectivity—has been crucial for future-proofing the city against more significant changes in the economy, population, and environment.

Green infrastructure in Yarrabilba is integrated into its main systems rather than being handled as a distinct planning category. Walking and bicycling routes are in line with green loops that lessen dependency on automobiles while promoting wellbeing and access to nature; landscape corridors function as both ecological connectors and networks for public mobility; and stormwater management is in line with the provision of open spaces. The goal is to build a community that coexists with and is part of the landscape, not merely parks or reserves.

Incorporating resilience thinking—not only to present environmental circumstances but also to potential hazards—was another goal of the planning process. The structure plan included climate adaption features, such as permeable surfaces, tree canopy targets, and heat reduction and shade providing. Crucially, these goals were incorporated into the regulated development requirements rather than being left to the discretion of individual developers or homeowners, guaranteeing uniformity and accountability throughout the whole delivery process.

A multi-layered sense of place results from the interaction of housing demand, environmental regulations, and infrastructural cooperation. Yarrabilba is developing into an urban ecosystem rather than merely a housing estate or greenfield neighbourhood. Both the demands of regional expansion and the obligations of ecological stewardship are embraced by its underlying logic. The project establishes a standard for how suburban growth might be rethought for a resilient and liveable future by prioritizing nature in its development and rejecting to view sustainability as a limitation.

In this regard, Yarrabilba's tale is about strategically planning for expansion rather than just accommodating it. It provides a crucial lesson: it is feasible to satisfy present demands without endangering the well-being of future generations when planning is based on ecological knowledge and long-term infrastructure thinking.

15.2 Designing the Urban-Ecological Interface: Built Form, Open Space, and Environmental Logic

The relationship between ecological integrity and urban development is not just an edge condition at Yarrabilba; rather, it is a guiding principle woven throughout the entire planning process. By designing a development pattern that values cohabitation over dominance, the project has sought to eliminate the historical dichotomy between nature and suburbia from the beginning. Here, environmental systems, open space, and constructed form are seen of as co-evolving layers that produce an enhanced urban experience in addition to a sustainable footprint.

Yarrabilba's terrain serves as the foundation for its spatial logic. In order to guide the location of roads, housing, public areas, and utilities, the design takes into account the topography, river flows, and biodiversity corridors that are now in existence. The plan leverages environmental thresholds as design signals and follows the land's contours rather than enforcing a strict grid. Instead of becoming barriers to be removed or pushed to the outskirts, this strategy guarantees that the site's ecological values—riparian corridors, native flora, and wildlife habitats—are preserved and highlighted as structural aspects of the urban form.

The way that Yarrabilba redefined open space is among its most noteworthy contributions to Australian urbanism. Here, green infrastructure plays a crucial role in the area's performance, identity, and liveability rather than being viewed as leftover acreage or aesthetic enhancement (Cheshmehzangi 2020). A vast network of green corridors has been established by connecting linear parks, retention basins, and native bushlands. These green corridors serve a variety of purposes, including stormwater management, ecological interconnectedness, microclimate regulation, and recreational enjoyment. Walking and cycling routes can now traverse through shaded, environmentally thriving terrain rather than around it because to these areas' active, navigable, and integrated design.

By aiming for a more compact and walkable development pattern without compromising the goals of suburban houses, the built form strategy enhances this environmental integration. Townhouses, courtyard houses, and dual occupancy are examples of medium-density housing designs that are grouped around important amenities, enabling higher population thresholds to sustain local business and public transportation. Affordability, diversity, and environmental performance have all been carefully balanced in the typologies and lot configurations. Orientation, ventilation, and solar access are examples of passive design elements that are promoted, and the materials used show a practical yet robust answer to Southeast Queensland's climate.

The public realm design of Yarrabilba is essential to tying the precinct's ecological and social goals together. With their narrow carriageways, shaded walkways, and numerous crossings to encourage pedestrian traffic, streets are intended to be social places. The distribution of civic areas and pocket parks throughout the neighbourhood guarantees that all inhabitants, irrespective of housing style, may easily access and enjoy them. Crucially, these areas are not generic; rather, they are created with ecological purpose and landscape character, frequently utilizing native plants, swales, and interpretative features that encourage locals to interact with the surrounding natural systems.

At the neighbourhood level, development sequencing has made it possible to supply green space and infrastructure either before or alongside housing, which is essential for creating an early sense of place and community identity. Green spaces are guaranteed to be lived rather than just promised thanks to its integrated delivery. Major open space corridors are paired with schools, community centers, and sports fields to create multipurpose hubs that support ecological health and anchor neighbourhood life.

The precinct's dedication to a new urban-ecological paradigm is particularly evident in its edge conditions. Through progressive density and green buffers, Yarrabilba's outer bounds integrate into conservation zones, agricultural fields, and riparian ecosystems instead of having distinct boundaries or development pods. By establishing distinctive thresholds that serve as community assets, these transitions lessen the boundary between development and the surrounding natural systems and provide both physical access to nature and a sense of psychological embeddedness within a larger environment.

Most significantly, Yarrabilba shows how environmental design promotes urbanity rather than hinders it. When nature is carefully incorporated and placed in key locations, it enhances everyday rhythms. It encourages casual get-togethers, promotes biodiversity, reduces heat, and increases resistance to climate extremes. By doing this, it questions conventional wisdom regarding suburban development and presents an alternative paradigm in which ecological intelligence is ingrained in urban planning and sustainability is perceived as a lived, everyday reality rather than a technical add-on.

The lessons learned from Yarrabilba are becoming more and more important as Australia continues to urbanize quickly, particularly in areas like Logan. They demonstrate that new communities that are socially and ecologically vibrant can be created with vision, dedication to design, and cross-sector cooperation (Queensland Chief Scientist 2011). In these communities, the city's boundaries signify the start of a more integrated urban future rather than the boundaries of urban ambition.

15.3 Everyday Urbanism: Placemaking, Participation, and Social Pulse

At Yarrabilba, the story of urban life is not only told through streets and buildings, but through the everyday rituals, interactions, and relationships that animate its spaces. The project has included a placemaking agenda from the beginning that aims to develop a resilient and socially diverse community rather than just a functional suburb. Early activation, community involvement, and social infrastructure have been prioritized in this master-planned setting just as much as the architectural design, laying the groundwork for a robust civic pulse that expands in tandem with the population.

Yarrabilba's dedication to establishing a feeling of place long before the last brick is built sets it apart from many other expansive greenfield developments. A unified character for the developing city has been shaped in large part by early investments in social programming, temporary centers, and community development officers (Cheshmehzangi 2016). Residents frequently discover a civic rhythm already established when they move into new residences, including playgroups gathering in the neighbourhood park, neighbourhood cookouts, sporting events, and seasonal celebrations that serve to unite individuals and establish a sense of belonging.

The provision of integrated community facilities is at the heart of this strategy. These are essential elements of the master plan, not extras; they are incorporated within walking distances and are made to be multifunctional, adaptable, and accessible. The community hubs in Yarrabilba are places for education, artistic expression, cross-cultural interaction, and helpful assistance. Operating in collaboration with neighbourhood organizations, they provide services like parenting classes, digital literacy initiatives, language instruction, and job training. By giving people skills and encouraging trust, belonging, and local networks, they contribute to the tangible development of social capital.

The social landscape of the town is also significantly shaped by schools and education areas. Co-locating youth programs, elementary and secondary schools, and early childhood centers has produced organic gathering spots that strengthen a feeling of community and sustained investment in the area. Within the larger precinct, parents, teachers, students, and service providers create micro-communities that enhance neighbourhood life and strengthen intergenerational links.

Public programming and events have been thoughtfully chosen to represent the variety of locals and to encourage inclusivity from the very beginning. Events at Yarrabilba are intended to be socially significant, place-specific, and interactive, whether they take the form of innovative art installations, health and wellness programs, or multicultural festivals. These brief interventions create moments that transform empty spaces into cherished ones by enlivening public areas and promoting communal memory.

Crucially, Yarrabilba views placemaking as an ongoing, dynamic process rather than a marketing gimmick. The development team has set up procedures for resident participation, feedback, and co-creation in coordination with local stakeholders. Residents are actively involved in shaping the precinct's future identity rather than being passive beneficiaries of planning choices thanks to community reference groups, participatory design sessions, and digital engagement tools. By providing flexible answers to the community's evolving needs and goals, this bottom-up participation enhances the top-down vision.

These societal goals are supported by the built environment. With their broad walkways, shady spots, and pocket parks that encourage casual conversation, streetscapes are made for more than simply movement. Town centers, which are already becoming important hubs for lifestyle, combine communal uses, health care, retail, and cafés to create a fine-grained, daily urbanism that encourages spontaneity and social ease. This combination of programming and design creates a vibrant social fabric that acknowledges that in a new suburb, belonging is not something that just happens; rather, it is developed through deliberate tactics. In a place where people arrive from different backgrounds, drawn by affordability, amenity, or lifestyle change, the role of public life in building identity is magnified. The social pulse of Yarrabilba is being actively written in the present rather than being passed down from the past.

Yarrabilba is a convincing example of how social sustainability can be ingrained in a new urban neighbourhood in this way. Early activation, inclusive placemaking, and participatory governance are given top priority, ensuring that people live a place with agency, connection, and care rather than just occupying it. The daily dance of life that turns a master plan into a vibrant, living community is the essence of everyday urbanism.

15.4 Lessons for Tomorrow's Suburbs: Equity, Adaptability, and the Long View

One of Australia's most important urban issues is how to expand cities without making the same pitfalls as previous suburban expansions. Yarrabilba offers a progressive answer to this problem. Its narrative of ambition, experimentation, and adaptation provides important insights into how to match the values of long-term site stewardship, equity, and resilience with large-scale development.

The project's dedication to inclusion—not just in the social sense, but also in the way it incorporates equity into the design process itself—is among its most significant accomplishments. Yarrabilba's strategy emphasizes the need for opportunity-rich environments for all demographics, from early investments in social infrastructure and relationships with local organizations to affordable housing programs and easily accessible public amenities. It questions the notion that high-quality urban living is only found in affluent neighbourhoods or inner cities. Rather, it demonstrates that if it is given priority from the beginning, a layered, diverse, and inclusive society may exist even in greenfield settings.

The emphasis on adaptability is equally central. Yarrabilba is a framework intended to change throughout time rather than a set plan. Whether it's staging infrastructure in alignment with growth, introducing new community facilities in response to demand, or using adaptive land-use arrangements to support future economic shifts, the precinct is structured around resilience. Being future-ready means accepting change as an inevitable aspect of city life, not just surviving shocks. The ability to make course corrections without losing coherence is crucial in a world where demographics, technology, and the climate are all changing.

An important factor in this adaptation is environmental integration. Yarrabilba is an example of how green infrastructure may improve a place's character and identity while fulfilling a variety of ecological, recreational, social, and climatic purposes. It illustrates how the conventional dichotomy between "urban" and "natural" is giving way to a more mutually reinforcing connection in which stewardship and growth support one another (Burton 2015). In the suburbs of the future, where urban expansion must be more restrained while yet providing housing and amenities on a large scale, this balance is crucial.

The way Yarrabilba frames suburban development as a communal endeavour rather than merely a market-driven one is arguably the most instructive. It poses the question of what it means to design for daily dignity, care, and belonging. It acknowledges that the true success of a location is determined by the quality of life that develops over time rather than just by the infrastructure provided or the number of houses constructed. This long-term perspective is uncommon but essential, particularly in a planning culture that prioritizes immediate results above long-term effects.

Of course, there are obstacles in the way. Careful consideration must be given to preserving quality as the precinct grows, preventing social stratification, making sure infrastructure keeps up with expansion, and integrating local governance systems that can support the community when the developer leaves. However, the fundamental choices taken at Yarrabilba provide a solid starting point for addressing these issues. Finally, Yarrabilba serves as a model for a distinct form of suburb: one that is ecologically conscious, community-oriented, and forward-thinking. It serves as a reminder that suburbs don't have to be unsustainable, expanding, or alienating. With determination, creativity, and perseverance, they may be hubs of meaning, connection, and resiliency—where the energy of the city flows through the suburbs rather than against them.

References

Alam K, Rahman A, Ranaweera K, Markar S (2014, January) Establishment of a hydrological benchmark for mitigating urbanisation impacts at Logan village, South-east Queensland. In: hydrology and water resources symposium 2014. Barton, ACT, Engineers Australia, pp 774–781

Bajracharya B, Hastings P (2020) Stakeholder engagement for disaster management in master-planned communities. Aust J Emerg Manag, 35(3):41–47

Burton P (2015) The Australian good life: the fraying of a suburban template. Built Environment 41(4):504–518

Cheshmehzangi A (2016) Temporary and temporality: public realm regeneration through temporary events. J Urban Regen & Renew 10(1):58–72

Cheshmehzangi A (2020) Reflecting on urban identities. In: Identity of cities and city of identities. Springer, Singapore, pp 133–178

Delfin Lend Lease (2004) Yarrabilba community strategy. Delfin Lend Lease, Brisbane

Logan City Council (n.d.) Yarrabilba, https://www.logan.qld.gov.au/about-council/about-the-city-of-logan/suburbs/yarrabilba

Queensland Chief Scientist (2011) Innovating queensland communities, document developed in November 2011, https://www.chiefscientist.qld.gov.au/__data/assets/pdf_file/0009/50004/innovating-qld-communities-2014-12.pdf

Queensland Government (2025) Yarrabilba priority development area: development scheme, 56 pages document https://www.edq.qld.gov.au/__data/assets/pdf_file/0022/99022/Yarrabilba-PDA-Development-Scheme-amendment-no.-2-2025-02-28.pdf

Urban Land Development Authority (2010) Yarrabilba urban development area: interim land use plan, https://www.statedevelopment.qld.gov.au/__data/assets/pdf_file/0015/13227/yarrabilba-interim-land-use-plan.pdf

Chapter 16
Nightcliff Foreshore Revitalisation: Coastal Urbanism for Climate, Culture, and Community in Darwin

16.1 Edge of the City, Heart of the Community: Reimagining Nightcliff Foreshore

One of the key initiatives in transforming Darwin's coastline into a public, ecological, and symbolic asset is the Nightcliff Foreshore Revitalization (Bowden 2014). This intervention is rooted in an existing location with strong local significance, cultural memory, and environmental fragility, in contrast to initiatives that arise from high-density aspirations or greenfield ambition. Nightcliff's shoreline has always provided a casual, understated link between people and place, and locals have long loved it for its picturesque cliffs, shady walking paths, and expansive vistas of the Timor Sea. However, the area was at risk due to aged infrastructure, coastal erosion, and climate stresses, making the revitalisation project urgently needed (Purwati 2001; Tonyes 2018). The result is a meticulous balancing act between ecological stewardship and public access, between continuity and regeneration, and between preservation and enhancement.

The Nightcliff foreshore, which is only ten minutes from Darwin's downtown, is more than just a place for recreation (Allman 2021). It is an integral part of the city's daily rhythms, acting as a linear park, a gathering spot, a path for fitness, and a haven from the tropical heat of the area. However, the cumulative effects of storm surges, sea level rise, and slow infrastructure deterioration made this edge more vulnerable over time. The realization that the coast needed to be redesigned rather than just defended influenced the urban strategy that underpinned the revitalization. The initiative aimed to improve the community's access to, use of, and maintenance of this area while integrating resilience (Williams et al. 2020) into the environment.

The revitalization initiative was intended to be a series of little, deliberate actions that add up to a more significant civic change rather than a single, spectacular gesture. Now, a new multipurpose route runs along the beachfront, enhancing connectivity without detracting from the site's natural charm (City of Darwin 2024). Water fountains, shade structures, seating areas, and educational signage all emerge in a rhythmic

A. Cheshmehzangi, *Australia's Urbanity*, Urban Sustainability,
https://doi.org/10.1007/978-981-95-3881-2_16

manner, encouraging rest, introspection, and learning. The project's fundamental philosophy—that public space is about more than just design, but about how design enhances everyday life—is reinforced by these little actions.

The enhancement of coastal edge management and erosion control has been one of the project's most obvious effects. Where loose rock walls and deteriorated boardwalks were once the predominant ad hoc solutions, the revitalized waterfront today has integrated landscaping and engineered interventions that replicate natural processes. Improved drainage, dune rehabilitation, and mangrove replanting are all components of a broader strategy to work with nature rather than against it. In addition to improving the foreshore's biological integrity, this has helped stabilize it.

Crucially, the project's success rests not only on improved facilities or environmental enhancements but also in the way it has strengthened the sense of civic responsibility. Early in the process, locals and community organizations were consulted on everything from signage preferences and cultural heritage allusions to accessibility issues. Their involvement contributed to the creation of a public space that seems welcoming and intimate. Through weekend markets, kid-friendly activities, community gatherings, and informal social contact, the foreshore has evolved into a space where local identity (Cheshmehzangi 2023) is expressed rather than enforced. The revitalized precinct has significance beyond its physical interventions thanks to this multi-layered use.

Additionally, Nightcliff Foreshore questions the widely held belief that coastal infrastructure ought to be protective or hard-edged. Rather, the initiative demonstrates how social and environmental goals may be served by a gentle, integrated approach. It provides a contrast to the conventional waterfront promenades found in many other Australian cities by accomplishing this. Here, the revitalization is expressed locally and is based on regional culture and landscape logic. Darwin's tropical ecology is reflected in the material palette, native plants, and climate sensitivity, which also allow for comfort, longevity, and continuity across time.

Nightcliff is a reproducible paradigm for how urban margins might develop with care as the city deals with growing population development and climate volatility. It serves as a reminder that revitalisation can sometimes be achieved by recalibration of intent rather than scale change. The Nightcliff Foreshore can be improved by listening, learning, and layering what already there. Revitalization is successful as a hub of civic spirit as well as infrastructure or amenity. It provides a durable urban edge that is both useful and significant by reestablishing people's connections to nature, community, and memory.

This research demonstrates how little coastal changes can have a significant civic impact within Darwin's larger urban environment. It redefines what it means to create urban resilience in the face of uncertainty by transforming the edge into a hub for local life. The foreshore is no longer just where the city meets the sea, it is where the city finds its pulse.

16.1.1 Coastal Challenges, Civic Opportunities

The revitalization of Nightcliff Foreshore resulted from the interlocking crises of urban neglect and climate vulnerability, not just from the wish to enhance a cherished public area. Nightcliff, like many northern Australian coastal margins, is becoming threatened by environmental change, particularly storm surge, sea level rise, and coastal erosion. At the same time, the area was underutilized, dispersed, and in some places hazardous due to years of haphazard urban growth and little investment. A rare opportunity to turn an endangered coastal strip into a robust, lively, and inclusive public space that both reflects and strengthens Darwin's urban identity was created by these twin pressures—environmental and civic.

The Nightcliff Foreshore is situated in an extremely exposed area. The location, which protrudes into the Timor Sea, is influenced by tropical monsoon patterns, cyclonic weather systems, and strong tidal movement. Coastal retreat has escalated in recent decades, exposing cliffs to greater degradation, harming vegetation, and destroying pedestrian trails. Although there had been isolated applications of hard technical solutions, these solutions frequently lacked integration with more general ecological or community concerns. Despite its popularity, the beachfront had devolved into a patchwork of shoddy facilities, disjointed walkways, and uneven infrastructure. Climate change intensified the urgency of a new approach, one that acknowledged the site's natural volatility while reinforcing its long-term resilience.

Simultaneously, urban characteristics produced a unique set of difficulties. Nightcliff's coastline edge has not been properly incorporated into the city's larger development concept, despite being a popular local destination. It struggled to satisfy the expanding needs of a diverse community, had poor signage and facilities, and lacked strong connections to nearby neighbourhoods. Darwin's ongoing growth and densification in strategic locations raised the importance of functional public areas. In this light, the untapped potential of the foreshore started to garner fresh interest, not only as a location to save but also as a community resource that was just waiting to be reclaimed.

This realization sparked a change in perspective, moving away from reactive repair and toward a more deliberate approach to revitalization. Instead of being a hindrance, the site's environmental limits served as a catalyst for innovation. Planners, designers, and community leaders started to rethink the foreshore as a place that might set an example for inclusive design and climate adaptation, rather than just as a landscape that needed to be preserved from the sea. The goal was to create something that spoke to Darwin's character, its climate realities, and its community aspirations—not simply to keep the line, but to enhance it.

The idea that public enjoyment and environmental preservation were complimentary rather than antagonistic goals was crucial to this shift. Aesthetics and usefulness were taken into consideration while designing erosion control measures; paths were rerouted and raised to protect trees and views; and native vegetation was not only preserved but also highlighted. Hard edges were included into seating or sculpture where they were necessary for protection. Instead of being engineered away, the natural drama of the shoreline was embraced. In order to promote active use and civic pride, new gathering places were also added, including playgrounds, art nodes, and shaded pavilions. Additionally, the project capitalized on momentum and timeliness. Particularly after local climate disasters and more extensive awareness efforts, there was a greater desire among the public for outdoor enjoyment and climate-conscious planning (Engage Darwin 2023). The Nightcliff revitalization was framed by the Darwin City Council and other stakeholders as an indication of the city's future direction, where community life, ecology, and infrastructure all come together, rather than as a standalone improvement.

Furthermore, the revitalization process was presented as a time for public engagement. In order to establish priorities, resolve conflicts between access and preservation, and make sure the project reflected local identity, community involvement was essential (Cheshmehzangi et al. 2023). The site was rediscovered from within, not reinvented from above. Local companies, Indigenous voices, longtime users, and more recent arrivals all helped to create a vision that felt both forward-thinking and grounded. By doing this, Nightcliff evolved from a planning project into a collaborative act of caregiving that is grounded in location and motivated by the aspiration for a better future for all. Lastly, Nightcliff Foreshore's revitalization shows how towns can intentionally spend to turn coastal issues into civic opportunities. A place where natural systems, public needs, and design thinking come together to create a more living and meaningful urban environment, what was once a vulnerable edge has now become a model for urban resilience.

16.2 Shaping the Resilient Coast: Landscape Logic, Material Design, and Public Access

The Nightcliff Foreshore Revitalization is an urban design project based on the delicate balance between enjoyment and resilience, pleasure and protection, rather than just beautifying a section of the coast. The project's success rests in its capacity to apply design thinking to an ecological challenge, transforming a strained coastline into a public space that is both environmentally sensitive and socially generous in an area that is becoming more and more exposed to the effects of climate change. This section examines how the thoughtful balancing act of landscape logic, material durability, and inclusive access has moulded the Nightcliff foreshore into a resilient municipal landscape.

Fundamentally, the revitalization proposal respects the natural topography of the shoreline. The design accepts erosion and tidal changes as forces to work with rather than against, embracing the coastline's changing shape rather than fighting it. As a result, a landscape-led design approach has emerged that views the coast as a narrative and a threshold. To embed its interventions within the site's natural logics, the project's spatial layout was driven by contour mapping, stormwater movement, wind exposure, and tidal flux. Every design choice reflects a thorough understanding of the topography and climate, from lounging places nestled inside vegetated buffers to winding pathways that follow the dune lines.

A robust material strategy that was created with Darwin's tropical climate and coastal characteristics in mind complements this site sensitivity. While seating and shelter components are built from weather-hardy metals and timbers that require little upkeep, hard surfaces are composed of textured, non-slip materials that can tolerate heat and salt exposure. Because the lighting is muted and protected, there is less ecological disturbance, especially for the local wildlife that moves through the area at night. In addition to lessening the impact on the environment, the use of sturdy, low-maintenance, and regionally relevant materials reflects the practical aesthetics of a city accustomed to the extremes of the rainy and dry seasons.

The new foreshore's accessibility is one of its distinguishing characteristics. Instead of a straight boardwalk, the idea offers several points of entry, halt, and movement. The purpose of these spatial rhythms is to promote rest, exercise, informal interactions, and intergenerational use. People of all abilities can enjoy both lofty views and beach-level activities thanks to gently graded ramps and expanded sidewalks. The path's nodes, which include drinking fountains, shaded pavilions, and interpretative information, encourage both group activities and solitary contemplation, reaffirming the foreshore's status as a public area accessible to all.

Crucially, the revitalized foreshore unites environmental infrastructure and public space rather than separating them. Incorporating water-sensitive urban design ideas with the user experience, drainage swales serve as green corridors. In addition to providing shelter from storm surges, raised berms also act as wind breaks and observation platforms. The landscape can serve as a live interface between urban life and coastal systems thanks to planting palettes that are chosen for their native resilience, seasonal variation, and ecological benefits. The end effect is a position where function becomes experienced, and form follows function.

The community's lived relationship with the coastline is reflected in the architecture, which goes beyond technological resilience and accessibility. The foreshore serves as a place for daily gatherings, including school excursions, weekend markets, sunset rituals, and times of quiet, in addition to being a place for strolling and bicycling. By improving the site's usability and ambience without overpowering its casual, relaxed vibe, the revitalization has accentuated this local culture. The arrangement of the furniture promotes staying rather than merely passing through. Lookout spots promote taking a minute to stop, think, or speak. Incorporated into the pavement or along the borders of the seats, public art features honour Nightcliff's Indigenous heritage and local history while adding nuanced layers of significance without overpowering the scene.

Importantly, shade has been viewed as a necessary infrastructure rather than an afterthought. Providing thermal comfort is essential for use in Darwin's extreme heat. To create microclimates that entice people to stop rather than just pass by, the revitalization includes extensive tree planting, light-framed shelters, and deliberate orientation to capture breezes. By extending the foreshore's usability beyond the early morning and late evening hours, these interventions support the public realm's status as a place that is habitable on a daily basis.

The Nightcliff Foreshore Revitalization's architectural success is largely due to its conservatism. It stitches together pieces of social and environmental infrastructure to create a coherent and readable whole rather than imposing a major architectural gesture. Instead of over-programming, it chooses lifespan and adaptability. It permits future modifications, seasonal changes, and emerging uses—all characteristics of a genuinely resilient public place.

Nightcliff offers a new standard for how coastal towns might consider their borders by emphasizing ecology, inclusion, and climate adaptability. It transforms fragility into vitality, resulting in a more resilient coastline and a more cohesive, liveable urban periphery. Through design, the foreshore has evolved into one of the city's most vital and expressive municipal spaces, in addition to serving as its perimeter.

16.3 A Place for All Seasons: Activation, Culture, and Everyday Belonging

Nightcliff Foreshore's revitalization is a reinvention of public life at the city's edge, going far beyond simple physical repairs. The foreshore has been thoughtfully activated to encourage not only sporadic visits but also consistent, daily interaction because it is a location that must adapt to both climate and community. It is now a location of familiarity and encounter, a place where Darwin's coastal culture is lived, shared, and redefined, thanks to careful placemaking and programming.

The fact that the foreshore now supports a changing rhythm of activity is among the most striking results of the revitalization. Runners and bikers frequently follow the covered trails before dawn, while others start their days with yoga, walking clubs, or peaceful contemplation by the lake. Families bring picnics in the late afternoon, kids frolic in the tide pools, and photographers and painters come in the golden hour. In the evening, the foreshore becomes a place where people congregate peacefully, whether they are couples watching the sunset, friends getting together at a food truck, or musicians playing a little concert as part of a community gathering.

This range of uses is not coincidental; rather, it is a result of a multi-layered placemaking approach that takes into account the climatic, cultural, and temporal aspects of public space. Instead of depending exclusively on planned activities, the foreshore has been planned to encourage unplanned usage and unofficial occupation. Pocket seating places provide for peaceful seclusion, while grassy clearings encourage unstructured play and group picnics. Even during warmest times of the day, visitors can stay comfortably thanks to shade structures and airy pavilions. The lighting design maintains the setting's low-key, intimate feel while promoting safe evening access.

An additional element of liveliness is added by events and cultural activations. Events like storytelling evenings, art fairs, community markets, and exercise classes occasionally liven up the area without overpowering its organic charm. Cultural heritage days, dry season festivals, and kid-friendly nature-based learning programs are examples of seasonal programming that is carefully chosen to fit local trends. A sense of authenticity and ownership over the area is ensured by these locally based activations, which are frequently co-created with artists, community organizations, and Indigenous caretakers.

The foreshore's identity is subtly but significantly shaped by the presence and knowledge of Indigenous people. The strong cultural ties to land and water are recognized through naming customs, integrated art forms, and interpretive signage. These actions, which are part of the larger story of inclusivity and shared stewardship, are not overt declarations but rather thoughtfully incorporated symbols of respect and acknowledgment. This strengthens Nightcliff's function as a gathering spot for the community and a cultural landscape while also giving everyone who uses the area a stronger sense of belonging.

The foreshore's accessibility to all populations is what makes it genuinely inclusive. Families utilize it as a place to connect and have fun, young people congregate on open lawns or shaded seats, and older folks use it for leisurely walks and socializing. The design's ability to provide a variety of experiences, i.e., from active to passive, lonely to social, guarantees that people will return not only for events but also because it is a location where everyday life can happen comfortably and easily. It is both a stage and a retreat, private and public.

This experience still revolves around the natural environment. The location's identity-forming and health-promoting features include views of the sea, exposure to the breeze, and closeness to coastal flora. For many people, going to Nightcliff is about grounding and wellbeing more than it is about social interaction. The emotional and cultural impact of the foreshore comes from this connection to nature. It is a place where people congregate and feel rooted; identity is established by enduring relationships with the land, water, and one another rather than merely architecture.

What distinguishes Nightcliff Foreshore is its emotional resonance and feeling of seasonal flexibility. It is a place that subtly welcomes use rather than demanding it. It allows for the convergence and coexistence of various lives, narratives, and experiences without ever coming across as unduly regimented or formal. It demonstrates how public areas may transcend utility and become meaningful places when they are created with empathy, humility, and ecological understanding. Ultimately, the revitalized waterfront promotes a kind of genuinely human everyday urbanism. It provides a space for relaxation, play, celebration, and introspection. It serves as a reminder that placemaking is essential to fostering civic identity and social resilience in cities impacted by climate change. In every season, Nightcliff serves as a serene yet potent illustration of what it means to be an urban resident at the water's edge.

16.4 Living with the Tides: Lessons for Adaptive Urban Edges

More than just a coastline improvement, Nightcliff Foreshore serves as an example of how towns may peacefully adjust to environmental change while enhancing the social and cultural lifeblood of the community. Nightcliff is a realistic and contextually aware example of how to coexist with, rather than resist, natural rhythms in an era when ecological degradation, heat stress, and sea level rise are upending traditional urban planning. It anchors Darwin's identity while providing more general lessons for Australian urbanism by transforming a fragile coastal edge into a durable public area.

Designing with the environment, not over it, is one of the most important lessons learned from the Nightcliff project. The goal of the revitalization was not to remove the coast's natural textures or harden it with concrete obstacles. Rather, it used a hybrid approach that combines accessible public infrastructure, natural planting, and erosion management. This ecological awareness improves usefulness rather than detracts from it. Through the use of porous materials, curved walkways, and coastal plants, the design creates a dynamic, dynamic, and safe environment.

This strategy is especially pertinent to other Australian urban and regional centers that are dealing with comparable pressures along their riverbank or coastal borders. Nightcliff shows that public amenity and climate adaption don't have to be mutually exclusive. These objectives can support one another with appropriate design thinking and community involvement, creating environments that are both generative and protective. In order to facilitate cultural and social activities without overpowering the scenery, the foreshore demonstrates how landscape-based initiatives may soften infrastructure.

The project's scope and limitations teach us another lesson. Being a showpiece development is not what Nightcliff aims to be. Its modesty—its ability to complement rather than alter the current landscape—is what gives it worth. In a city like Darwin, where extensive, expensive urban initiatives may seem divorced from local realities, this philosophy is especially potent. Nightcliff's success stems from its awareness of people and place. By promoting local stewardship as opposed to passive consumerism, it fosters a more leisurely and personal interaction with the city.

The project's focus on continuity, i.e., between the past and present, culture and climate, memory and aspiration, is equally significant. Building on Nightcliff's past as a communal area, the revitalized foreshore maintains its informality while enhancing its usefulness. It is not a brand-new location; rather, it has been revitalized by its users. Fostering urban identity requires this continuity, particularly in a place like Darwin where narratives are frequently dominated by transience and flux.

The importance of regular public space as a type of civic infrastructure is shown by Nightcliff's social and cultural relevance. People look for locations that provide comfort, familiarity, and shared significance in the face of fast change. One such location is Nightcliff, which is truly lived and cherished, not because it was promoted as legendary. Although its boundaries may change with the tides, its location at the center of the city never changes. Nightcliff Foreshore provides an optimistic and realistic lesson as coastal cities around Australia get ready for the unpredictable tides of climate, demographics, and economic change. It serves as a reminder that adaptive urban margins present chances to rethink how cities might interact with the natural world, culture, and society, rather than merely posing technical difficulties. Nightcliff shows us how to design with vision, humility, and care by living with the tides.

References

Allman K (2021) Travel and lifestyle: Darwin and surrounds. LSJ: Law Soc J, (80), 56–58

Bowden M (2014) Long-grassers seen as blight on Darwin's iconic foreshore. Eureka Street 24(23):19–20

Cheshmehzangi A (2023) Regenerating the local characteristics: the impacts on city life experience and identity of Shangri-La's old town. In: mapping urban regeneration: city life experiences in Yunnan, China. Springer Nature Singapore, pp 219–237

Cheshmehzangi A, Zou T, Tang T, Xie L (2023) Exploring influential factors for decision-making in urban development: urban heritage protection and displacing living communities. Bhumi, Plan Res J, 10(1)

City of Darwin (2024) Strategic and priority projects plan 2030, https://www.darwin.nt.gov.au/sites/default/files/publications/attachments/105883StrategicandPriorityProjectsPlan2030Report_FA(1).pdf

Engage Darwin (2023) Nightcliff village masterplan, https://engage.darwin.nt.gov.au/nightcliff-village-masterplan

Purwati P (2001) Reproduction in a Fissiparous Holothurian, Holothuria Leucospilota Clark 1920 (Echnodermata: Holouroidea) in tropical waters of Darwin, northern territory. Charles Darwin University (Australia), Australia

Tonyes SG (2018) Sand dynamics in Darwin Harbour: a tool for coastal management. Doctoral dissertation, Charles Darwin University (Australia)

Williams R, Meharg S, Muster TH (2020) Tracking Darwin-A framework for. Part or in whole and any information or material contained in it

Part IV
Industrial and Economic Innovation Precincts

Developments that drive economic growth, employment, and high-tech industries:

- **Citiswich Business Park, Ipswich**—Queensland's largest industrial and logistics hub.
- **Aerotropolis Western Sydney**—A proposed cutting-edge economic precinct centered around the new Western Sydney Airport.
- **Lot Fourteen, Adelaide**—A high-tech innovation precinct fostering digital and defence industries.
- **Fishermans Bend, Melbourne**—A transformation of former industrial land into a future-forward mixed-use district.
- **Macquarie Point Urban Renewal, Hobart**—A waterfront innovation hub driving economic and research growth in Tasmania.

Chapter 17
Engines of the New Economy: Citiswich Business Park and the Urbanisation of Industry in Bundamba, Queensland

17.1 The Industrial City Reimagined: Citiswich and the Pulse of Regional Growth

The concept of industrial development in the twenty-first century has been boldly recalculated by Citiswich Business Park. This master-planned district, which is situated in Southeast Queensland on the outskirts of Ipswich, deviates from the traditional perception of an industrial estate as a separate and peripheral land use (Citiwich, n.d.). Rather, Citiswich is a prime example of a more intentional urbanism—an employment ecosystem where scale, design, infrastructure, and accessibility all work together to promote local economic development. It is a strategic intervention in Ipswich's transformation from a post-industrial regional city (The Industrialist 2025) into a vibrant node within the state's larger metropolitan constellation, not just a logistics hub or industrial zone.

With an area of more than 350 hectares, Citiswich is the largest industrial and logistics park in Queensland in terms of footprint (Bennett and Bennett n.d.). It serves as the foundation for a new economic geography that goes hand in hand with the changing dynamics of urban development in the state. Ipswich is now being positioned as a vital logistics and employment corridor, whereas previously regional centers were reduced to satellite roles with diminishing industrial activity and limited growth potential (Worrall et al. 2004). Situated near Brisbane's main freight lines, the Ipswich Motorway, and the convergence of the Warrego and Cunningham Highways, Citiswich takes advantage of its strategic transportation links to cater to local, interstate, and export markets. Its placement is no coincidence; it is the result of years of planning around transportation-oriented economic growth and represents a more general change in the way jobs, land, and infrastructure are being integrated for growth.

Citiswich appears at a moment when Australian towns are reframing the significance of industrial zones as facilitators of municipal resilience and balance rather

A. Cheshmehzangi, *Australia's Urbanity*, Urban Sustainability,
https://doi.org/10.1007/978-981-95-3881-2_17

than only as containers of economic function. Industrial land is now a vital component of urban development, not an afterthought, given the rising land values, housing demands, and a renewed appreciation for domestic manufacturing and supply chain sovereignty. In response to this backdrop, the business park provides a versatile land-use platform that can house a range of tenants, including light manufacturing, commercial offices, service-based sectors, and logistics behemoths and warehouses. Scalability is crucial, and large-format lots are combined with built-to-suit development facilities that let companies grow over time within a unified spatial framework.

Citiswich's development also represents a significant turning point in Ipswich's urban development. The city, which was formerly characterized by its history of rail and mining, is now establishing itself as a major force behind the decentralized urban future of Southeast Queensland. In this way, Citiswich is not a stand-alone development; rather, it is an economic engine that is intricately linked to the regional planning concept. Employment hubs like Citiswich serve as a vital counterpoint to the significant residential growth occurring in Ipswich's neighboring greenfield districts, such as Springfield, Ripley, and Redbank. This is indicative of a more comprehensive planning philosophy that aims to localize economic activity inside sub-metropolitan nodes, co-locate housing and employment, and lessen the burden of commuting.

The extent to which Citiswich has been master-planned for adaptation and integration in addition to function sets it apart from previous models of industrial districts. The business park was designed with a comprehensive infrastructural backbone, including utility corridors, high-capacity highways, and stormwater systems, to anticipate long-term growth, as opposed to a piecemeal approach. By using this strategy, Citiswich has been able to draw in large amounts of private investment and prestigious tenants while maintaining the precinct's ability to adapt to changing market demands. In a period characterized by changing technology, climate imperatives, and supply chain complexity, the development's planning logic gave equal weight to short-term deliverability and long-term resilience.

Crucially, the Citiswich story is not solely about square footage or economic indicators. It also involves a nuanced rethinking of the ways in which industrial development supports civic aspirations and urban identity (Cheshmehzangi 2023). Through employment, infrastructure, and place logic, Citiswich contributes to determining the pulse of urban life, although lacking the pedestrian-friendly scale of typical neighbourhoods and the symbolic weight of a public plaza. It is a part of a larger movement to acknowledge that urbanity is found in everyday landscapes of work, logistics, and manufacturing as well as in places like cafes, high streets, and waterfronts. In this sense, Citiswich broadens the definition of what it means to become a city in modern Australia.

Citiswich Business Park is a key component of the urban future of the area as Ipswich continues its swift change. It highlights the changing function of industrial precincts as deliberate components of a more dispersed, resilient, and inclusive metropolitan structure, in addition to their position as economic hubs. Through master-planned delivery, infrastructure-led investment, and strategic land use, Citiswich demonstrates how industrial urbanity can be made readable, linked, and vibrant.

17.1.1 Planning for Industry, Designing for Urban Futures

Citiswich Business Park's planning is based on a conscious balancing act between spatial strategy and economic needs. Citiswich was designed from the ground up as a high-performance logistics and industrial precinct with a specific purpose, in contrast to many typical industrial zones that developed gradually or as reactive land uses. A regional vision that recognized the changing economic landscape of South East Queensland—one in which regional employment creation, decentralized growth, and freight efficiency would emerge as crucial pillars of long-term sustainability—guided its planning.

A collection of integrated land-use planning techniques that gave priority to future scalability, strategic position, and land availability are at the core of Citiswich's development. Because of the site's level terrain, close proximity to major highways, and distance from sensitive environmental zones, the Ipswich City Council, in collaboration with the Queensland Government and private developers, saw it as having great potential early on. The idea of a next-generation industrial hub that could serve large-format users and gradually adjust to new industry demands was greatly aided by this positioning.

The land-use zoning of the property was designed to accommodate a variety of commercial and industrial uses, ranging from logistics and light industry to warehousing and business park operations. Instead of being dispersed at random, these uses were included into a larger master plan that prioritized tenant clustering, site efficiency, and traffic circulation. Major logistics operators were given access to large, contiguous pieces of land, while service-oriented enterprises were given smaller lots that offered flexibility and resilience to changes in the market. By providing a range of lot sizes and built-form alternatives, Citiswich was able to accommodate various phases of business expansion and draw in a diverse tenant base.

Importantly, infrastructure-led development was also considered when planning Citiswich. The precinct's deployment was accelerated by the early installation of trunk infrastructure, including high-capacity internal highways, sewer and water connections, and stormwater systems. These investments demonstrated confidence in the precinct's long-term viability and appeal to investors, and they went beyond simple technical provisions. Crucially, a key factor in the masterplanning process was transportation connection. Improvements to the nearby road networks, along with the site's direct access to the Warrego and Cunningham Highways, allowed for easy freight transit between the park and the Port of Brisbane, the Brisbane Airport, and other regional destinations. Citiswich's importance in Queensland's larger economic geography was strengthened by its ability to establish itself as a state-scale distribution node thanks to this logistical advantage.

Another important factor in determining Citiswich's strategic orientation was economic forecasting. With the western corridor expected to see substantial population growth and Southeast Queensland emerging as one of the fastest-growing regions in the nation, policymakers clearly focused on fostering local employment opportunities that could lower outbound commuting and improve economic self-containment. In response to this requirement, Citiswich created a plan for development that would create jobs and increase with the city's residential population. It was master-planned as an integrated part of the metropolitan area, where economic function is interwoven into the spatial logic of city-building, rather than as a standalone estate.

Regional growth frameworks and employment plans that saw industrial precincts as more than merely back-end areas helped to support the planning process. Both the Ipswich Planning Scheme and the Southeast Queensland Regional Plan of the Queensland Government acknowledged the necessity of high-employment nodes to counterbalance population growth. Citiswich was able to go from concept to delivery with clarity and coordination because to the supporting planning environment this policy superstructure created.

Citiswich stands out for transforming industrial planning into an urban thought process. The district has been envisioned as an urban ecosystem with economic resilience at its heart, rather than just a collection of loading docks and warehouses. A design concept that anticipates growth rather than responds to it is demonstrated by the inclusion of landscaping buffers, internal roadways built for high-load vehicles and plans for future integration of public transportation. Lastly, Citiswich's planning narrative unveils a new paradigm for industrial urbanism: one in which economic insight, spatial intelligence, and long-term land-use visioning come together to build not just a business park but also the urban future's foundation. Citiswich is a prime example of how designing for industry can also entail designing for urban vitality, accessibility, and resilience as Australia works to create more sustainable, productive, and balanced metropolitan areas.

17.2 Infrastructure Meets Innovation: Spatial Strategy, Access, and Scalability

An example of infrastructure-led innovation is Citiswich Business Park, where industrial expansion is now a well-balanced engine of regional progress rather than an afterthought. Its strategic coordination of accessibility, scalability, and flexibility forms the basis of its spatial logic, each of which supports its position as a pivotal project for Ipswich's economic future. In contrast to the traditional perception of remote industrial estates, Citiswich is strategically and physically integrated with the larger metropolitan and transportation networks of Southeast Queensland. As a result, the precinct serves not only the enterprises inside its borders but also the regional systems that rely on and are fed by it. It is both pragmatic and performative.

Citiswich's access strategy, which is based on its location at the intersection of three important thoroughfares—the Ipswich Motorway, the Cunningham Highway, and the Warrego Highway—is the key to its success. The precinct has direct access to Brisbane, the Port of Brisbane, the larger Ipswich area, and the southern and western corridors of the state thanks to this triangle, which creates a high-capacity freight and commuter loop. The site was chosen as a node in the state's broader plan for decentralized economic activity that relieves strain on urban cores and unleashes regional potential, so it wasn't a coincidence. Citiswich offers operational fluidity for logistics, distribution, and light manufacturing businesses by blending in seamlessly with this transport spine.

Beyond the highways, internal road planning within the park demonstrates an attention to flexibility and flow. In addition to being functional, wide, truck-compatible streets are built to be future-proof, able to adapt to changes in last-mile logistics requirements, delivery technology, and freight modalities. By employing design signals to strengthen legibility and orientation, the street grid avoids the obscurity that characterizes many industrial estates. A degree of spatial care typically found in commercial or municipal districts is shown by clear sightlines, layered planting, and navigational infrastructure. Both the space's users and the precinct's identity are enhanced by these design choices.

The scalability of Citiswich's infrastructure is essential to its development. The concept was initially designed to support a variety of building styles, lot sizes, and operating needs. Because of this flexibility, tenants from small logistics start-ups to national distribution centers have been able to co-locate inside the same larger framework, resulting in a dynamic industry ecosystem rather than a static mono-use zone. The installation of digital, water, and power infrastructure included redundancy and upgrade capabilities in addition to meeting immediate needs. Such planning has allowed the precinct to adjust to technological advancements and regulatory changes without requiring a complete rebuild in an era of changing energy requirements and sustainability standards.

This model of coordinated regional development avoids the pitfalls of industrial sprawl and enables long-term urban coherence. Crucially, Citiswich is not isolated; rather, it is part of a purposefully planned industrial corridor that connects to allied zones like the Swanbank Enterprise Area and the Ripley Valley development to the south. This networked logic allows for the distribution of roles across the industrial geography of Ipswich, where Citiswich serves as the logistics interface, Ripley as a spine of residential and mixed-use, and Swanbank as a future hub for green-energy innovation.

Citiswich's scalability extends beyond its physical growth to include its economic flexibility. Precincts like Citiswich are evaluated not only on their capacity but also on their potential to change course as consumer behaviour, international trade patterns, and automation continue to alter industry needs. Businesses can quickly adapt to changes in the market thanks to the park's modular design, strong connectivity to fiber-optic networks, and intelligent utilities. Because of its adaptability, Citiswich is positioned as a living economic engine that can ride the tides of change while maintaining its position as a hub for local employment and regional output.

Finally, Citiswich's spatial techniques also subtly but significantly influence how people view industry. The precinct challenges the antiquated idea that industrial zones must be visibly and functionally separated from the rest of the city by improving the quality of the entering experience, landscaping, and infrastructure. This is important for attracting new jobs in the future, coexisting with the growth of nearby residential areas, and creating a favourable image of the Ipswich area as a center of innovation and industry in the twenty-first century.

In conclusion, Citiswich's ability to combine economic vision with spatial performance is what makes it successful. It illustrates how industrial urbanism, when properly designed and integrated, may help redefine a developing region's space, economy, and symbolism in addition to improving logistics efficiency. Its infrastructure is not merely concealed or buried; rather, it forms the basis of a new urban logic that views the industrial sector as essential to Australia's modern urban development rather than as a peripheral one.

17.3 The Urbanity of Industry: Employment, Movement, and Everyday Experience

Citiswich Business Park is a thriving urban ecosystem where work, logistics, and mobility come together to create a multi-layered experience of daily industrial life (Ipswich City Council 2019). It is more than just a neighborhood of trucks and warehouses. As Australia's industrial precincts move past its one-purpose histories, Citiswich is a prime example of a modern form where urban sensibility and economic intensity coexist. In this instance, the rhythms of freight operations, shift changes, and delivery timetables do not exclude the site from city life; rather, they create a new form of urbanity that is grounded in function but flexible and accessible to design.

The scale of employment created by Citiswich is substantial. The park serves as a regional anchor for jobs in warehousing, light manufacturing, distribution, and related professional services, and it is expected to employ tens of thousands of people. Drawing from Ipswich, Brisbane's western corridor, and new suburbs in the Ripley and Springfield growth zones, the employment base is varied not only in terms of sector but also in terms of shift patterns and demographics. The park's urban logic depends on this workforce integration since it makes Citiswich a destination rather than just a location of employment and connects the precinct to the bustling residential neighbourhoods nearby.

Industrial precincts depend on movement, and Citiswich was created to support it on several levels. Wide arterial loops and service roads intended for effective turning, docking, and stacking are traversed by heavy vehicles. A finer layer of human-scale access—bike links, shuttle potential, and pathways—complements these logistical movements and responds to employees' daily commutes. Citiswich is a subtle but significant change from industrial zones, which historically prioritized vehicles over people. This transition acknowledges that industrial workforces, like all others, should have access to high-quality mobility networks, readable settings, and areas for respite and comfort.

This leads us to the changing urbanity of the actual workspaces. Industrial architecture in Citiswich is rapidly responding to new demands about flexibility, design quality, and well-being in addition to scale and function. When possible, breakout areas, end-of-trip amenities, and natural light are being incorporated into the architecture of large-format warehouses. In order to create layered interfaces between operational and professional zones, offices and admin pods are integrated into warehouse volumes with glazing and green edges. In a competitive job market, this blending of typologies promotes the retention of a talented personnel and helps to create a more dynamic working environment.

This change is also starting to show in the precinct's support services. Recognizing that an industrial workforce requires more than simply loading docks to operate efficiently, Citiswich incorporates space for food shops, service stations, and possible daycare and training centers. By transforming the precinct from a production site to a regular workplace, these service nodes aid in humanizing the area. Additionally, they offer the kinds of transitional spaces that promote informal contact and workplace culture, such as lunch locations, meeting rooms, and casual gathering places.

The shift to a hybrid economic model from mono-functional zoning is another feature that sets Citiswich apart. Logistics, clean tech, construction supply chains, and even light advanced manufacturing are among the many business kinds that the park is designed to support. A degree of innovation and cross-sector synergy that is not achievable in older, more strictly zoned industrial estates is made feasible by this diversity, which also strengthens the local economy. Citiswich is a major hub in a new type of urban industrial fabric that is more vibrant, more mixed, and more interwoven into the city's daily life as the larger Ipswich area develops into a metropolitan sub-center.

The park questions preconceived notions about what industrial regions are and can be in a number of ways. Even if trucks and forklifts continue to play a major role, they are a component of a much broader ecosystem that also involves daily routines, human movement, economic shifts, and spatial quality. The patterns of lunch breaks, walking routes, informal contacts, and site circulation that transform the industrial into the daily are just as much a part of the workplace here as shift bells and deliveries.

Citiswich serves as a reminder that industry's urbanity is a progression rather than a conflict. Precincts like Citiswich demonstrate how employment hubs may enhance not only GDP and freight capacity but also the lived experience of urban life as Australia attempts to reorient its economic geography for a more distributed and sustainable future. By doing thus, they contribute to redefining the lines between the beginning of labour and the end of cities.

17.4 Designing Tomorrow's Economic Landscape: Lessons from Citiswich

A fascinating look into the future of industrial urbanism is provided by Citiswich Business Park, where jobs, logistics, infrastructure, and local identity come together to form a dynamic and intentional economic environment (Cheshmehzangi and Willman 2023). Precincts like Citiswich offer concrete examples of how industrial development can be transformed from its conventional form into something more resilient, strategic, and place-responsive as cities throughout Australia struggle with the demands of decentralization, housing supply, and employment distribution.

Fundamentally, Citiswich exemplifies the importance of adaptability and long-term planning. Because of its size (more than 350 hectares) and long build-out horizon, the precinct was never intended for transactional development or short-term turnover. Rather, Citiswich was intended to be a phased, flexible ecosystem that could adapt to changing market trends, ranging from conventional distribution and warehousing to sophisticated manufacturing and the circular economy sectors. Tenants may adjust when logistical systems, environmental regulations, and technology change because to its adaptable land parcels, modular infrastructure, and zoning permits.

In the industrial setting, environmental performance is also being redefined. Although high-impact footprints are frequently linked to industrial complexes, Citiswich incorporates sustainable design concepts at the precinct level. Effective road alignments, thoughtful planting corridors, and stormwater management all contribute to the site's overall environmental impact management. As demands for ecological integration and carbon reduction increase, Citiswich offers a paradigm where environmental stewardship is ingrained in the precinct's operations as a design requirement rather than an afterthought.

Most importantly, the park illustrates how employment clustering supports local vibrancy. Through the consolidation of several business kinds, including SMEs, service operators, and national logistics suppliers, Citiswich facilitates the creation of an employment zone that can support itself. This employment concentration creates a strong economic catchment for Ipswich and the larger Southeast Queensland corridor by supporting auxiliary services, training initiatives, and transportation connections. The occupations created here contribute to the need for housing, education, mobility networks, and the dynamics of metropolitan growth in general.

Perhaps the most lasting takeaway from Citiswich, however, is the importance of design—not just in the aesthetic sense, but also in the intentional structuring of identity, legibility, and functionality. Citiswich has adopted urban design principles to provide useable green buffers, clear wayfinding, considerate interface zones, and visual coherence on a large scale in an area that is frequently dominated by utility. These actions represent a turn in industrial design toward dignity, where the operator's, visitor's, and worker's experiences are all incorporated into the urban story. This in turn promotes long-term economic sustainability, investor attraction, and branding.

Citiswich contributes to the reinterpretation of industrial land in the twenty-first century in a number of ways. Industrial precincts like this one, which were once marginalized as unseen production zones, are now becoming important contributors to the economic geography of Australia by creating jobs, bolstering regional development, and sustaining logistical networks. They are now hubs for interaction, adaptation, and development rather than merely throughput. The task for cities as they develop will be to replicate what Citiswich has done first: create economic areas that are not merely efficient but long-lasting, and not simply productive but also purposeful. Here, on the brink but unstoppable, is where the industrial city of the future starts; on the edge, but firmly in motion.

References

Bennett and Bennett (n.d.) Citiswich Industrial Park. https://bennettandbennett.com.au/projects/citiswich-industrial-park

Cheshmehzangi A (2023) Regenerating local industries and urban experiences that matter. In: Mapping urban regeneration: city life experiences in Yunnan, China. Springer Nature Singapore, Singapore, pp 23–75

Cheshmehzangi A, Willman E (2023) The city as palimpsest to redefine the role of architectural continuity in re-establishing urban identity in post-industrial landscapes. J Urban Regener Renew 16(3):340–357

Citiwich (n.d.) Citiwich Business Park. https://www.citiswich.com.au/

Ipswich City Council (2019) Statement of proposals: Draft Ipswich Planning Scheme. https://www.ipswich.qld.gov.au/__data/assets/pdf_file/0014/113207/Statement_of_Proposals.pdf

The Industrialist (2025) Strong demand as Ipswich business hub nears completion—JLL & Knight Frank. https://www.theindustrialist.com.au/properties/2025/06/25/strong-demand-ipswich-business-hub-nears-completion-jll-knight-frank

Worrall R, Neil D, Grigg A, Brereton D (2004) Legacy mining areas on the urban fringe: a study of Ipswich, Queensland. In: JKMRC international student conference, Vol 5, Brisbane, Luettu, p 2014. https://www.csrm.uq.edu.au/publications/legacy-mining-areas-on-the-urban-fringe-a-studyof-ipswich-queensland

Chapter 18
Aerotropolis Rising: Reimagining Economic Urbanism at the Crossroads of Mobility, Innovation, and Place in Western Sydney

18.1 Airborne Ambitions: The Aerotropolis as a New Urban Model

One of the most ambitious urban initiatives in modern Australian history is the Western Sydney Aerotropolis, a daring fusion of economic strategy, spatial design, and infrastructure (Jazbec et al. 2020). In contrast to other airport-centric schemes, the Aerotropolis is intended to be a fully integrated urban ecosystem centered around the Western Sydney International (Nancy-Bird Walton) Airport, rather than just a transportation hub. The Aerotropolis, which has its roots firmly established in Greater Sydney's Three Cities Strategy, is a reinvention of city-making, innovation, and regional equity for the twenty-first century. It is more than just a logistics hub or employment cluster.

Situated in the center of Western Parkland City, the Aerotropolis holds a prominent position in Sydney's urban story. Sydney's eastward-skewed economic and physical expansion for many years exacerbated disparities between the eastern and western suburbs in terms of infrastructure delivery, service quality, and employment access. As a counterweight to the established economic weight of the Eastern Harbour City, the Aerotropolis anchors expansion in Sydney's west, unlocking potential in a hitherto underdeveloped region. Its introduction is a symbolic and economic gesture toward a future that is more diffused and polycentric, in addition to being a spatial correction.

The combination of local activation and global aspiration is what sets the Aerotropolis apart. Although the airport itself is anticipated to become a significant aviation gateway and sustain tens of thousands of employments, its true power is found in its surroundings (Dench 2024). The precinct, which spans over 11,000 hectares of strategically located employment land, is designed to support a variety of high-value industries, including advanced manufacturing, aerospace, defence, agro,

A. Cheshmehzangi, *Australia's Urbanity*, Urban Sustainability,
https://doi.org/10.1007/978-981-95-3881-2_18

logistics, and research. The outcome is an economic diversification model that repositions Western Sydney as a global leader in next-generation productivity rather than as a satellite.

The air, terrestrial, digital, and institutional logic of interconnectedness forms the framework of this redesigned metropolitan paradigm. The precinct makes use of its close proximity to several multimodal transportation options, such as new highways, freight routes, and a dedicated metro line that will connect the Aerotropolis to the public transportation network in Greater Sydney (Hawken 2021). Connectivity is more than just physical, though. The Aerotropolis is creating memoranda of understanding with international universities, research institutes, and multinational corporations through governance coordination and international partnerships—sowing the knowledge and cooperation ecosystem needed for the innovation economy to flourish.

The timing of the Aerotropolis is also important. The Aerotropolis is being positioned as a testbed for innovative industries, renewable technologies, and resilient infrastructure as Australia deals with the economic changes of a post-carbon future. Underpinned by a dedication to performance-based planning, it provides a platform for the development of carbon–neutral logistics, circular waste systems, energy-efficient buildings, and adaptive urban design. Although these goals are still primarily aspirational at this point, the development's size and greenfield status offer a unique chance to test concepts that would be challenging to implement in more established urban regions.

Importantly, the Aerotropolis is not being imagined as a standalone commercial complex (Barker Ryan Stewart, n.d.). Mixed-use neighbourhoods, educational precincts, cultural institutions, and environmental corridors are all included in the larger planning goal. To provide a new paradigm of hybrid urbanity that reflects how people actually experience cities today, it aims to obfuscate the conventional boundaries between work, learning, living, and enjoyment. In this way, the airport connects many spatial and social functions into a coherent urban framework, acting as a point of convergence rather than a border.

But there are difficulties. The success of the precinct depends on decades of consistent investment, integrated governance at several governmental levels, and sincere interaction with the surrounding populations, many of which are displaced or severely disrupted. Protections for affordability, accessibility, and long-term viability must be paired with the promise of jobs and change. The Aerotropolis runs the risk of turning into a dispersed area of privileged industry (Wastnage 2019) rather than a welcoming urban hub if it is not properly managed.

The Aerotropolis' goal is still to establish a new type of city, one in which innovation clusters do not arise at the expense of community, infrastructure leads but does not dominate, and spatial equity becomes a guiding principle rather than a retroactive adjustment. As the ground gets ready to take off, the Aerotropolis challenges Australia to reconsider what a city in the twenty-first century can be if it is designed with the sky, the network, and the human experience in mind rather than the automobile or the office. The Western Sydney Aerotropolis is a novel approach to urban and economic development in its audacity, scope, and integrated reasoning. It suggests that the airport of the future is a location where futures arrive rather than only a place to go.

18.1.1 From Greenfield to Global Gateway: Strategic Vision and Spatial Opportunity

The Western Sydney Aerotropolis is the result of one of the biggest greenfield opportunities in Australia's history of urban development (NSW Government, n.d.)—a point in time when economic ambition, massive infrastructure delivery, and long-term strategic planning come together to create a global gateway. The Western Sydney International (Nancy-Bird Walton) Airport, a catalyst project that anchors a vision for a new urban economy and spatial transformation in Sydney's west, lies at the center of this opportunity. However, the Aerotropolis is not just about the construction; it is also about the planning, the people it will serve, and the futures it will open up.

This change is largely being driven by the airport itself. It is positioned as more than just a transit hub because of its advantageous location in the center of Western Parkland City, one of the three metropolitan centers listed in Greater Sydney's Three Cities Strategy. It is the catalyst for a whole ecosystem of innovation, jobs, and mixed-use growth. The airport serves as the hub of the Aerotropolis Core Precinct, bringing together planning initiatives in the surrounding area, accelerating infrastructure delivery, and stimulating investment. Additionally, it supports more general national objectives related to export-led economic growth, decentralization, and sovereign industrial capabilities.

The Aerotropolis's spatial logic shows a conscious change in planning goals from reactive zoning to proactive place-making. Authorities have positioned the region as a future metropolis in and of itself, one that combines economic productivity with environmental sustainability and human-centered urban form, rather than treating it as a generic logistics zone or a remnant industrial hinterland. In order to integrate sophisticated manufacturing, agribusiness, education, health, defence, and digital technology into a cohesive urban framework, a network of interconnected precincts is intended to be established.

Equally important are the governing frameworks that make this goal possible. The Western Parkland City Authority offers a cross-jurisdictional mechanism for coordinating goals, funding infrastructure, and guaranteeing implementation amongst agencies and stakeholders. It is backed by the governments of New South Wales and Australia through the Western Sydney City Deal. Eight local councils have joined forces under the City Deal model to work together on common goals like enhancing liveability, creating jobs, and developing smart city frameworks. Coordination of the many tiers of economic policy, transportation service, and spatial development necessary to realize the Aerotropolis vision (Searle and O'Connor 2022) depends on this whole-of-government approach.

Precinct-based planning is a key component of this governance approach. The Aerotropolis is organized into discrete precincts, each having its own unique land use, industry focus, and urban design, as opposed to embracing a single land-use plan. These include the Agribusiness Precinct, Northern Gateway, Aerotropolis Core, and the adjacent environmental and employment zones. This strategy enables gradual development that adapts to environmental restrictions, infrastructural delivery, and market demand. Additionally, it makes it possible for institutions and industries with similar goals to group together, creating synergies that transform the precinct (Morrison and Van Den Nouwelant 2020) from a business park to an innovation corridor.

This strategic stance has been made possible in large part by investments in transportation. High-frequency rail access between the Aerotropolis and important nodes like St Marys and the larger metropolitan network will be made possible by the planned Sydney Metro—Western Sydney Airport line. The dedication to accessibility, sustainability, and less reliance on personal vehicles is furthered by this integration with public transportation. Additionally, it bridges the historical gap between the eastern central business district and the western suburbs, symbolically and practically linking Western Sydney to the city's economic core.

The Aerotropolis' scope of ambition to strike a balance between local benefit and global integration sets it apart from other significant initiatives. The precinct is intended to connect to global innovation networks through memorandums of understanding with research institutions, aerospace businesses, and international universities. In order to guarantee that the economic advantages reach the surrounding communities, many of which have long suffered from underinvestment and restricted access to employment, job targets, training initiatives, and local procurement strategies are being established concurrently.

An further generational opportunity to incorporate environmental stewardship into extensive urban expansion is provided by the Aerotropolis. Initiatives like biodiversity offsets, green infrastructure corridors, sustainable water management, and net-zero building commitments show a goal to develop an urban environment that is resilient, healthy, and liveable in addition to being a viable employment zone (Roggema 2024). Planners and designers have the unique opportunity to include sustainability into the precinct's design from the beginning rather than attempting to retrofit it decades later because so much of the area is still undeveloped.

In conclusion, Australian urban policy is evolving and increasingly recognizing the importance of integrated planning, cross-sector collaboration, and precinct-scaled visioning, as seen by the Western Sydney Aerotropolis. It illustrates how a single piece of infrastructure, e.g., the airport, can spark a far more extensive change in how cities expand, connect, and reshape their futures, in addition to changes in land use and economic function. This is not just a project for development. It is a fresh approach to envisioning what cities can become when they are designed as platforms for the potential of the future rather than as continuations of the past.

18.2 Urban Form and Economic Logic: Planning for Innovation, Scale, and Movement

The Western Sydney Aerotropolis' master planning poses a unique and challenging task: creating a completely new urban form that is both practical for the international industries it aims to draw in and flexible enough to meet the demands of coming generations. The Aerotropolis is a purpose-built vision that is developing on primarily undeveloped terrain, in contrast to many inherited urban places that develop gradually and naturally over time. This presents a once-in-a-lifetime opportunity, but also a great deal of responsibility, to integrate economic logic into spatial organization and make sure that innovation, place-making, and mobility are all intertwined from the start.

The logic of co-location is central to the Aerotropolis. Advanced manufacturing, aircraft, agribusiness, and defense are examples of industry clusters that are not just confined to discrete areas. Rather, in order to maximize innovation spillovers, minimize transportation bottlenecks, and promote information exchange, the strategy purposefully clusters comparable activity in close proximity. This clustering is made possible while maintaining spatial clarity by the precinct model, which is separated into function-specific areas such as the Aerotropolis Core, Northern Gateway, and Agribusiness Precinct. Green corridors, digital networks, and common public infrastructure bind these places together, ensuring that the Aerotropolis operates as an interconnected urban ecosystem rather than a collection of discrete zones.

In this sense, scale is closely related to urban form. The Aerotropolis is a city-forming endeavour rather than a neighbourhood. Its size—more than 11,000 hectares—allows it to accommodate a few business zones, civic infrastructure, logistics activities, and knowledge centers. Nevertheless, the design places a high priority on readability and navigation despite its scale, guaranteeing that residents, employees, and guests all experience a logical spatial logic. To balance regional access with fine-grained interior permeability, important roadways, public transportation lines, and movement corridors are organized around a hierarchical grid. This strategy is essential to advancing the precinct's goal of being both locally accessible and globally linked.

One fundamental design principle is movement. Smooth transportation both inside and outside the Aerotropolis has been a top priority for planners since the beginning. The precinct's ability to function as a logistical powerhouse is ensured by freight rail planning and arterial road renovations, while the new Metro line will serve as the spine for high-capacity public transportation. But mobility methods are not just about efficiency. Street design incorporates networks for cyclists and pedestrians, with a focus on end-of-trip amenities, green connections, and walkable centers. This represents a larger movement away from the car-dominated paradigms of old industrial estates and toward multi-modal, human-scaled spaces that promote accessible and healthy daily living.

The Aerotropolis' attempt to balance structure and flexibility is among its most striking features. The physical form is being designed to be adaptable throughout time, even while land parcels are set aside for specific businesses or purposes. Phased infrastructure delivery, flexible utility networks, and modular plot designs allow the precinct to adapt to shifting market needs and technology advancements. It is difficult to include this flexibility without compromising planning coherence or a distinct feeling of urban identity.

The spatial story revolves around research districts and innovation hotspots. To attract talent and investment to the area, university campuses, innovation hubs, and collaborative spaces are being positioned as the cornerstones of precinct life. The planning concept aims to integrate these institutions into a continuum—where knowledge generation, prototyping, and production take place within the same urban fabric—instead of perceiving them as distinct from industrial and commercial activity (Barker Ryan Stewart, n.d.). The Aerotropolis' post-industrial goal is to become a location where ideas are not only born but also realized, tested, and scaled, and its combination of education, industry, and urban life is an excellent example of that goal.

Additionally, mixed-use development is important, especially in the Aerotropolis Core. Here, businesses are envisioned coexisting with lodging, services, short-term lodging, and public facilities to create a district that operates day and night. These mixed-use nodes add vitality and human presence, which enhances placemaking and liveability, even if a large portion of the larger Aerotropolis is centered on jobs. They also aid in drawing in a larger workforce, which includes highly qualified individuals who want a certain level of convenience and lifestyle integration in their workplace.

A strategic shift from legacy industrial planning to next-generation economic urbanism is seen in the Aerotropolis' urban design initiatives. Effective land use is only one aspect of it; another is creating an atmosphere that fosters innovation, draws in a wide pool of people, promotes sustainability, and creates lasting economic benefit. In this sense, the Aerotropolis' spatial organization serves as both a canvas and a catalyst, influencing how people travel, interact, and create within the developing metropolis while housing industries of the future.

18.3 Urbanity at Altitude: Life, Work, and Place-Making in the Aerotropolis

Although the Western Sydney Aerotropolis is frequently discussed in terms of economic productivity, logistics, and infrastructure, its long-term viability will depend just as much on how well it functions as a lived environment. Here, the concept transcends commercial districts and industrial corridors to address a more profound urban issue: establishing an area where people live, work, socialize, and feel like they belong. The Aerotropolis is about creating a new kind of urbanity where everyday life, culture, and community thrive alongside business, not just about creating an economic engine.

Prioritizing human-scale planning within a large-scale framework is essential to achieving this goal. Despite the precinct's size, civic nodes—each intended to offer pedestrian access to necessary services, facilities, and gathering places—intersperse the masterplan. With the Aerotropolis Core evolving as the densest and most mixed-use center, these nodes are thoughtfully positioned throughout the precinct. In order to provide a strong sense of urban presence even within a newly constructed city-form, offices, educational institutions, short-term lodging, retail establishments, and dining venues are all planned to congregate in a single walkable neighborhood.

This new urban identity is woven together by networks of open space and green space (Cheshmehzangi 2020). The spatial arrangement incorporates parklands, urban woods, drainage corridors, and recreational pathways as guiding frameworks for development rather than as afterthoughts. Green infrastructure is purposefully included in the design to connect disparate land uses, improve walkability, and offer cooling in a climate-challenged setting. In particular, the South Creek Corridor serves as a linear public space and an ecological spine, promoting biodiversity and providing both locals and workers with recreational activities and active transportation options.

The precinct's dedication to liveability is further demonstrated by the incorporation of water infrastructure into the public space. Creating moments of connection between people and the landscape is a key component of water-sensitive urban design, which goes beyond stormwater management. Daylighted streams, bioswales, and wetlands serve as social and aesthetic anchors in addition to being useful infrastructure. In this way, water influences the Aerotropolis's public spaces' aesthetics and sensory value in addition to its environmental sustainability.

Another cornerstone of the Aerotropolis' human-centered design philosophy is mobility. Beyond the main transportation facilities, such as the Metro line and freight lines, a strong focus is placed on establishing an environment that is hospitable to cyclists and pedestrians. In important metropolitan locations, street design promotes slower vehicle speeds, while end-of-trip amenities, shaded pathways, and green corridors are designed to make walking and bicycling practical and desirable. High-frequency public transportation stops located within ten minutes' walk of employment and recreational areas demonstrate a strong commitment to the concepts of modern transit-oriented development.

Precinct planning gives significant consideration to social infrastructure. In order to provide the people who will live and work in the area with access to the basic necessities of daily life, schools, libraries, health services, childcare facilities, and community centers are being incorporated into each stage of development. In what will initially be a temporary and changing population, these services also contribute to the development of social cohesiveness and community identity (Cheshmehzangi 2014). Planning that prioritizes accessibility, diversity, and striking a balance between care and productivity helps to lessen the difficulty of establishing a feeling of place from the bottom up.

Most significantly, the Aerotropolis model acknowledges that a future city must support culture. Public art and cultural programming are suggested as essential elements of the precinct's activation plan. The objective is to cultivate a local culture that aligns with its First Nations background and its global economic agenda, ranging from event venues and pop-up markets to permanent installations and public storytelling. This is especially crucial in a district that is intended to accommodate foreign workers, tourists, and students—where urban vibrancy and cultural expression are essential to elevating the city above a mere business transaction.

The result is a precinct that aims to change the way people perceive industrial and economic hubs. In addition to housing cutting-edge enterprises, the Aerotropolis aims to ground them in a community that prioritizes daily life, civic engagement, and environmental well-being. In terms of aviation as well as the raising of design standards, public values, and long-term thinking, it is an urbanity at altitude. The Aerotropolis hopes to become a location where human experience and economic ambition are inextricably linked by integrating urban life into its design.

18.4 The Future in Formation: Lessons in Purpose-Led Urban Innovation

One of the most ambitious infrastructure-led urban experiments in contemporary Australia is the Western Sydney Aerotropolis, a project that aims to transform not just a location but also an urban and economic development paradigm. The Aerotropolis is fundamentally a forward-thinking plan that is influenced by concepts of innovation, connectedness, and national competitiveness. However, the precinct's success will ultimately hinge on how well it strikes a balance between these goals and more profound commitments to inclusion, equity, resilience, and place-making as it transitions from vision to implementation.

The fact that infrastructure by itself does not constitute a city is among the most important lessons to be learned from the Aerotropolis. The physical framework may be provided by freeways, airports, and logistical networks, but the spaces in between—civic engagement, social cohesiveness, and a clear comprehension of the human needs that underlie economic productivity, are what sustain urban life. In this sense, the Aerotropolis serves as a reminder that purpose-driven urban innovation needs to enter the architecture of belonging rather than just the infrastructure of expansion.

The project's size is both a strength and a risk. It has the ability to decentralize Sydney's economic geography, establish a high-performing node that draws both local talent and international industries, and open new job markets. However, if planning is not continuously inclusive and flexible, such large-scale interventions also run the risk of producing fragmented, uneven, or monofunctional results. Making sure that the advantages of the Aerotropolis—its employment opportunities, transit choices, housing options, and public areas—are fairly shared by Western Sydney's varied and expanding population is a significant problem.

A key component of this difficulty will be governance. The Aerotropolis necessitates an agile, transparent, and participatory governance approach due to its several levels of government, wide range of stakeholders, and intricate infrastructural interconnections. After the initial planning, coordination between the education, transportation, environment, housing, and industrial sectors must be maintained. Decisions on continued development must incorporate community input, not just at consultation stages. The construction of the Aerotropolis will take place across several developing phases, each of which calls for careful management and social responsibility.

High standards must also be applied to the precinct's environmental mission. Although ideas for water-sensitive architecture, green corridors, and sustainable mobility are encouraging, implementation and upkeep will be difficult, particularly in the face of market volatility and rapid development demands. The Aerotropolis can be used as a national model to future-proof large-scale development against environmental instability while improving ecological value and liveability in Australia, which is facing increasing climate concerns.

Most significantly, the Aerotropolis forces Australia to reconsider its definition of urban innovation. Innovation encompasses more than just digital infrastructure or high-tech industries. It also involves the invention of the social contract, policy, procedure, and spatial thinking. A city is considered truly creative if it actively reshapes the future in ways that are more sustainable and egalitarian, in addition to anticipating it. The Aerotropolis is, in many ways, the future in formation. From coastal concentrations to inland development, from infrastructure ambition to civic possibilities, it reflects the conflicts and aspirations of a country changing its urban boundaries. Its future won't be simple or straightforward. Its biggest contribution, however, may have been demonstrating how purpose-led development can result in meaningful urban living when combined with inclusive, grounded, and adaptable design.

References

Barker Ryan Stewart (n.d.) Western Sydney aerotropolis: Australia's newest and thriving commercial city. https://brs.com.au/western-sydney-aerotropolis-australias-newest-and-thriving-commercial city/

Cheshmehzangi A (2014) Spatial syntagma and identity of a place: sensing, relating to, and knowing a place. J Hum Behav Soc Environ 24(7):799–810

Cheshmehzangi A (2020) Reflecting on urban identities. In: Identity of cities and city of identities. Springer Singapore, Singapore, pp 133–178

Dench A (2024) Response to question on notice-Ally Dench-Committee on transport and infrustructure-Airport Aerotropolis. https://policycommons.net/artifacts/15882777/response-to-question-on-notice/16773637/

Hawken S (2021) Parkland visions: designing the Western Sydney aerotropolis. Landsc Archit Australia 169:22–27

Jazbec M, McGee C, James G (2020) Sustainability and heat report for Western Sydney aerotropolis. https://opus.lib.uts.edu.au/bitstream/10453/159754/2/0620018.10_Western_Sydney_Aerotropolis_Sustainability_and_Heat_Final_DRAFT_WITH_ALT_TEXT_21_October_2020.pdf

Morrison N, Van Den Nouwelant R (2020) Western Sydney's urban transformation: examining the governance arrangements driving forward the growth vision. Aust Plan 56(2):73–82
NSW Government (n.d.) Western Sydney aerotropolis. https://www.planning.nsw.gov.au/plans-for-your-area/priority-growth-areas-and-precincts/western-sydney-aerotropolis
Roggema R (2024) Climate resilient urban landscape in Western Parkland City, Sydney, Australia. https://academiccommons.columbia.edu/doi/10.7916/1dnd-j274/download
Searle G, O'Connor K (2022) Flawed vision? Sydney's three cities metropolitan strategy. Urban Policy Res 40(4):369–388
Wastnage J (2019) Bringing industry to airports: how the US experience can help Western Sydney to take off. https://apo.org.au/node/232156

Chapter 19
Lot Fourteen Business and Technology Precinct: Reprogramming Adelaide's Driving Force Through Innovation

19.1 Heritage to Horizon: Transforming Lot Fourteen with Innovation

Lot Fourteen is a powerful drama of urban renewal, as the forces of the past collide with the energy of emerging economies. This sector, which is situated on the site of the former Royal Adelaide Hospital, is neither a remote tech park on the outskirts nor a brand-new suburban campus (Cunningham and McCutcheon 2023). A central city redevelopment project that rewires Adelaide's very center for the twenty-first century is something far more ambitious. Adelaide is establishing itself as a globally significant player in the knowledge economy by cultivating a network of enterprises in artificial intelligence, space technology, defence innovation, and creative industries within the walls of its old sandstone building.

The realization that the abandoned hospital site, which took up an entire city block at the entrance to North Terrace's cultural boulevard, was too strategically significant to be reduced to just another piece of real estate is where Lot Fourteen's tale starts. State leadership purposefully chose to take the long view, keeping public ownership and repurposing the property as a municipal asset for innovation, rather than relying on private residential or commercial redevelopment (Cunningham and McCutcheon 2020). This signalled a turning point in urban thinking, moving away from seeing land largely as a commodity and toward seeing it as a tool for cultural and economic transformation.

Lot Fourteen's multi-layered identity (Cheshmehzangi 2025) is what gives it power. It serves as a national symbol for Australia's defence and space industries on one level. Another is that it is a walkable civic and cultural destination in a city that is renowned for its small size and robust institutional presence. In between, it serves as a venue for urban innovation, public involvement, and entrepreneurship. The precinct represents a unique fusion of technical aspirations and historical depth. Buildings that once housed operating theatres now contain start-ups and innovation labs. Originally used as roads, courtyards have been transformed into public

A. Cheshmehzangi, *Australia's Urbanity*, Urban Sustainability,
https://doi.org/10.1007/978-981-95-3881-2_19

gathering spots. Lot Fourteen's unique urbanity is defined by this architectural and symbolic harmony between the past and the future.

Importantly, the area was intended to be a location that is intricately woven into the rhythms and flows of Adelaide's central district rather than a closed innovation enclave (McIntyre et al. 2021). Its placement on North Terrace, surrounded by universities, important civic institutions, the Botanic Gardens, and the Art Gallery of South Australia, guarantees that it will always be permeable and readable by the general public. It is a civic precinct that has been purposefully included into the city's cultural and educational center (McIntyre et al. 2021) rather than a "tech bubble" cut off from its surroundings. Laneways, landscape corridors, and event spaces are examples of this openness, which is a distinguishing approach that attracts people and makes it difficult to distinguish between innovation, community, and culture.

Lot Fourteen is essentially a case study of how Australian cities might transform their institutional or post-industrial lands into areas of nation-building rather than just rehabilitation. Its contribution to Adelaide's shift from an industrial past to a future driven by information and innovation is what truly defines its success, not merely square metres or commercial worth. This shift has been supported by a distinct narrative that places an emphasis on civic significance, cross-sector cooperation, and long-term public results.

A whole-of-government strategy has been used to assist the precinct's growth, with both the federal and state governments making significant investments. However, it's not just a top-down narrative. Additionally, collaborations with academic institutions, start-ups, cultural organizations, and First Nations leaders have been necessary to activate Lot Fourteen as a vibrant space. These collaborations guarantee that innovation is viewed as a more comprehensive social and cultural process rather than being limited to technology terms. This notion is emphasized by the precinct's incorporation of the Aboriginal Art and Cultures Centre (Tarrkarri) (Woods Bagot n.d.), which highlights identity and cultural continuity as crucial elements of Australia's innovative future.

Additionally, Lot Fourteen is essential to rebalancing Adelaide's economic geography. The city is leveraging this district, which has historically been formed by jobs in manufacturing and government, to diversify and prepare its workforce for the future. The concept supports public transportation use, hospitality, housing demand, and inner-city vibrancy by bringing jobs, talent, and energy into the urban core by locating high-value enterprises in the center rather than in outlying areas. This is not merely spreadsheet-based planning; it is urban design-based economic growth.

In the end, Lot Fourteen is a daring urban experiment that acknowledges the city as a site for revitalization rather than a static shape. Its transformation from a nineteenth-century health campus to a twenty-first-century innovation ecosystem shows what is possible when cities invest with purpose, plan with vision, and design with care. Adelaide has done this in a way that allows heritage to be reinterpreted rather than erased by modernity, providing a powerful example of how Australian cities may integrate civic life, innovation, and history into a single, coherent urban framework.

19.1.1 Strategic Repositioning: Innovation Precincts and Economic Renewal

Lot Fourteen's transformation from a former hospital site into one of Australia's most well-known innovation areas was the result of a well-planned strategy driven by the confluence of political will, economic intent, and design vision rather than an unintended consequence of urban redevelopment. At the core of this change is a novel model of intergovernmental partnership that brought local, state, and federal priorities together around a common goal: to establish an economically significant node in Adelaide's city fabric that can spur a new wave of industrial diversification and civic innovation.

The former Royal Adelaide Hospital was situated on a site of great cultural and spatial significance on the eastern outskirts of Adelaide's central business area. It was a remarkable municipal asset, but one that ran the risk of becoming obsolete after the hospital was moved because of its close proximity to the Botanic Gardens, universities, and North Terrace cultural institutions. The South Australian Government came up with a daring idea to use the site as a catalyst for economic rebirth and global positioning, instead of dividing the property up for traditional redevelopment or letting it lie in a state of transitional limbo.

This choice was supported by a well-defined policy goal: to shift South Australia's economy away from its waning reliance on conventional manufacturing and toward high-value, knowledge-based industries. The state was seeking to strengthen this trajectory with innovation ecosystems that might support wider technological advancement and digital capabilities. The state had already started to draw significant investments in the defence sector, especially in the shipbuilding and submarine industries. Combining cutting-edge research, entrepreneurship, and cultural infrastructure into one walkable precinct, Lot Fourteen became the physical and metaphorical center of this change.

Both the Australian and South Australian governments firmly supported the precinct's official debut, jointly allocating funds for significant infrastructure improvements, industry attraction initiatives, and important anchor projects. Lot Fourteen was transformed from a real estate play into a collaborative national endeavour thanks to this interagency cooperation. With the announcement that the Australian Space Agency will establish its headquarters at Lot Fourteen, federal funding in particular provided crucial impetus. This immediately raised the precinct's prominence and attracted a global network of affiliated businesses and research collaborations.

In order to guarantee synergies and cross-pollination, the industry sectors of space, defence, artificial intelligence, machine learning, and the creative industries were strategically layered. These industries were chosen due to their potential for expansion as well as their compatibility with South Australia's current institutional framework and labour market goals. The site's reputation as a significant innovation zone was further cemented by the presence of important research institutions including the SmartSat Cooperative Research Centre and the Australian Institute for Machine Learning. Co-location was crucial; government agencies, corporates, researchers, and start-ups were purposefully grouped together to promote talent circulation, commercialization, and knowledge sharing.

However, Lot Fourteen's success has been shaped not only by the enterprises it houses but also by the way it has been incorporated into Adelaide's current public, cultural, and urban scene. A permeable and interconnected urban experience has been made possible by its proximity to North Terrace's establishments, including the University of Adelaide, the State Library, and the Art Gallery of South Australia. The precinct feels like a continuation of the city's civic spine rather than a place of isolation or artificiality because to pedestrian connections, planted public areas, and heritage buildings that have been conserved.

Cultural inclusivity and public accessibility have also been emphasized as being essential to the precinct's long-term identity. The goal of the planning was to establish a space for public interaction, learning, and innovation rather than a closed corporate campus. The Tarrkarri Aboriginal Cultural Centre, which will grow to be one of the most important First Nations institutions in the world, is the most notable example of this (Architecture, AU 2023). Its existence serves as a potent counterweight to the frequently exclusive inclinations of high-tech districts worldwide, reaffirming the need of cultural acknowledgment in innovation settings.

Lot Fourteen's staging and spatial layout have also been thoughtfully planned to strike a balance between ambition and adaptability. The area is being developed gradually rather than entirely at once, enabling flexible responses to community input, tenant demands, and market demand. Because of its adaptable structure, it has been able to stay resilient and relevant in the face of changing global circumstances while gradually assembling a community of workers, inventors, and city people who support its day-to-day operations.

Lot Fourteen is much more than just the result of sound planning; it is an example of what can be achieved when planning is centered on a common goal of civic engagement, economic revitalization, and spatial excellence. It serves as an example of how abandoned property in the center of a city may be developed into a nationally significant precinct, one that adds to the live narrative of a city that is changing with attention to detail, vision, and creativity, in addition to its economic figures.

19.2 Architecture of Innovation: Spaces, Infrastructure, and Civic Form

In addition to creating a high-performance innovation area, Lot Fourteen's architectural and urban design methods are essential to creating an approachable public space right in the middle of Adelaide. Fundamentally, Lot Fourteen is not a campus isolated from the city; rather, it is intentionally open, porous, and designed to facilitate mobility, interaction, and a variety of uses. This has been accomplished by carefully balancing the adaptive reuse of historic structures, the incorporation of contemporary design, and the development of multi-layered public spaces that foster social interaction and economic activity.

The site's history is deeply respected at the outset of the architectural approach. The stone structures of the former Royal Adelaide Hospital, which have long been a part of the city's institutional fabric, have been conserved, renovated, and repurposed for contemporary purposes. Their adaptive reuse provides a rich material continuity that grounds the precinct within Adelaide's architectural language. Instead of demolishing and beginning afresh, the precinct has a dual identity thanks to the integration of these heritage assets with progressive infrastructure: one that respects its history while expressing optimism for the future. A story of continuity and change is graphically told by the interaction of steel detailing, glass expansions, and sandstone walls; this is both conceptually and aesthetically in line with innovation itself.

The way that additional structures have been added to the site is equally significant. These buildings are audacious without being overbearing. They stay away from the fortress-like settings that are frequently observed in other international innovation areas, maintaining a mid-rise scale that honours both the human experience and the surrounding cityscape. Transparency, modularity, and ecological performance are highlighted in this light, open, and technologically advanced architectural language. The building shapes are oriented to maximize daylight, ventilation, and the interface with green spaces in order to create work environments that encourage welfare and collaboration.

A coherent and readable spatial network forms the precinct's core. The inside arrangement of Lot Fourteen has been designed with walkability and easy navigation as top priorities. The site is threaded through by a core spine that anchors movement across its different nodes and is energized by seating places, gathering spots, and planted components. The smooth transitions between the cultural institutions on North Terrace to the Botanic Gardens and the adjacent urban borders are ensured by this spine's connection to the larger city grid. In this approach, the design ties innovation to the city's daily life by fortifying both symbolic and physical ties.

Landscape is a structuring instrument, not an afterthought. The precinct's edges are softened, shade is provided in Adelaide's hot environment, and biodiversity is supported through the use of green infrastructure. Sustainable urban drainage techniques are used on-site to manage stormwater, and planting plans highlight native species and seasonal variations. In addition to providing chances for casual gatherings and passive enjoyment, these green layers interwoven between buildings and walkways enhance the feeling of place. They aid in communicating that this is a place of civic engagement and public life in addition to employment.

The material selections made by the precinct also reinforce its overarching theme of robustness, flexibility, and visibility. The predominant materials include high-quality, low-maintenance options like weathering steel, exposed concrete, repurposed wood, and perforated metal screens. These are employed to produce layered, tactile façades that are climatically responsive and aesthetically pleasing. In addition, transparent frontages and clear glazing encourage visual connection between interior and outdoor areas, fostering an air of openness and visibility for all parties.

Movement systems are purposefully made to be simple and easy to understand. Access for cyclists and pedestrians is given priority, while automobile traffic is kept to the periphery. The precinct is easily navigable by persons of various abilities and mobility choices because to its ample pathways, bike racks, and accessible entrances. Service vehicles are also subtly accommodated inside the network, preserving both utility and spatial quality.

The result of this spatial plan is a precinct that, in addition to housing industry, physically exemplifies the concepts of innovation. Lot Fourteen does not like the walled-off areas of corporate research parks or the tech enclaves of Silicon Valley. Rather, it creates a civic setting where landscape and architecture work together to promote transparency, inquiry, and cooperation. Lot Fourteen establishes a standard for how architecture and urban planning may transform innovation precincts from functional to cultural by fusing the old with the modern, indoors and out, formal and casual. It proves that areas of economic production can—and ought to—be hubs for innovation, community, and deep urban ties.

19.3 City Life, Culture, and Collaboration: Placemaking for Innovation Communities

The development of Lot Fourteen is profoundly cultural in nature rather than merely physical or economic. A strong dedication to placemaking as a means of encouraging creativity, teamwork, and public connection is at the core of the precinct's urban logic. The district is intended to be a vibrant, living component of Adelaide's civic and cultural character rather than a closed collection of offices. At this point, creativity transcends technical progress and turns into a communal and social phenomenon.

Lot Fourteen's philosophy is based on the idea that creativity flourishes in environments where people interact. In order to do this, the precinct's programming and design collaborate to create opportunities for cross-sector engagement between students, researchers, entrepreneurs, creatives, and the public. The site has a rhythm of activity thanks to a carefully planned calendar of open forums, networking events, cultural displays, and public presentations. These gatherings take place in a variety of public areas, including adaptable outdoor courtyards, casual meeting places, and inside multifunctional rooms. By fostering an atmosphere of openness and welcome, they make sure that the knowledge generated in the precinct engages the city and does not remain within.

Another essential element of the precinct's placemaking is cultural identity (Cheshmehzangi 2020). via active involvement with local communities and cultural practices, as well as via artwork and interpretation signage, the site acknowledges the history and presence of Aboriginal people. The precinct's spatial and symbolic language acknowledges and incorporates the Kaurna people's connection to the land. Lot Fourteen is anchored in place by this inclusion, which emphasizes the need for innovation to be connected to memory, context, and the history of the land it is situated on. The precinct turns into a place of respect and memory in addition to creating the future.

Strategic use of public art fosters a feeling of vitality and individuality. The precinct has a distinct visual language thanks to lighting interventions, interactive artworks, murals, and sculptures that bring life to the ordinary. These are integrated elements of the placemaking approach, not ornamental extras, and are meant to arouse interest, dialogue, and a sense of place. The incorporation of cultural clues into the architectural environment guarantees that Lot Fourteen is a location with personality, richness, and a feeling of community rather than a generic tech park.

Lot Fourteen's daily rhythms are meticulously orchestrated through a blend of official and casual public areas. The purpose of food and beverage establishments is to promote loitering, drawing employees, guests, and bystanders into a common social area. Shaded walkways, street-level activation techniques, and outdoor dining places all contribute to making sure the facility is active beyond business hours. The area accommodates a variety of employment patterns that contribute to its urban life, whether people are attending an evening event or just stopping by for coffee in between meetings.

Another important factor in fostering a collaborative culture is the merging of enterprise and education. The precinct houses university collaborations, research projects, start-up incubators, and defence innovation programs, allowing for everyday interactions between industries that might otherwise stay isolated. Open atriums, translucent façades, and shared workspaces promote visual connectedness and a group mentality. This method promotes interdisciplinary thinking as the rule rather than the exception and dismantles barriers across fields.

Lot Fourteen reinterprets what a workplace is in this way. It is more than just a place to do duties; it is a social organism where openness, experimentation, and interaction constantly shape and redefine culture. The precinct may act as a catalyst for both economic development and a new sort of urban culture based on inclusivity and creativity thanks to its diverse programming, public realm generosity, and civic ambition. The ability of Lot Fourteen to humanize innovation is ultimately what makes it successful. The precinct shows that innovation can be lived, shared, and celebrated in ways that benefit the city and the communities it serves by integrating the catalysts of economic growth into an environment that is approachable, lively, and culturally sensitive.

19.4 Lessons in Civic-Led Innovation: Precinct Thinking and the Post-Industrial City

An excellent example of how post-industrial Australian cities can embrace innovation without losing their sense of civic identity is Lot Fourteen. The precinct demonstrates that innovation may be based in place, people, and purpose rather than destroying legacy in favour of newness or separating industry from regular urban life. It is a type of precinct thinking that challenges us to think about how we construct things, socially, economically, and culturally, in addition to what we create.

Lot Fourteen's adoption of adaptive reuse is among its most illuminating features. The precinct reimagines the old Royal Adelaide Hospital complex rather than destroying it. Instead of being conserved as immobile monuments, heritage buildings are transformed into living infrastructure that houses public facilities, research facilities, co-working spaces, and creative studios. This strategy conveys a crucial lesson: tabula rasa development is not always necessary for innovation areas. In actuality, adding new uses to existing forms can result in more vibrant, richer urban settings that respect a city's history while preparing it for the future.

The leadership of the government plays an equally significant influence in determining the course of the precinct. Lot Fourteen is the result of intentional, long-term cooperation between the federal and state levels, supported by a compelling policy vision that prioritizes innovation as an economic and civic imperative. The stability and assurance required to draw in investment, form institutional alliances, and deliver infrastructure in line with long-term objectives are provided by this type of coordination. Lot Fourteen serves as an example of the strength of a shared vision and civic dedication in a time when dispersed governance frequently impedes urban progress.

The precinct's ability to adapt is another one of its strengths. Lot Fourteen was created as a flexible framework rather than a closed system, allowing it to adapt to changing community demands, industry trends, and new technological advancements. Public areas support a variety of programs, landscapes support a range of uses across time, and buildings are made to be repurposed. Given the speed at which cities and technology are changing, this degree of inherent resilience is crucial. Precincts that embrace creativity and reject rigidity are ones that prosper throughout time.

The notion that innovation precincts have to be exclusive or inward-looking is also contested by Lot Fourteen. The precinct creates a permeable contact with the rest of the city by integrating cultural venues, open spaces, and inclusive programming into its core. This has important ramifications for communities looking to draw in talent, hold onto creativity, and establish a reputation as hubs for people, not just businesses. The urban economy now encompasses more than just markets and infrastructure; it also involves experience, involvement, and a sense of belonging.

Fundamentally, Lot Fourteen represents a new form of economic geography that situates innovation in a larger framework of civic value. Creating a culture of curiosity, cooperation, and inclusivity that extends outside the precinct itself is more important than just creating jobs or financial gains. This concept reminds us that effective precinct planning must go beyond industry silos to support urban life in all its dimensions—economic, social, and cultural—and can be used as a model for other cities navigating post-industrial transitions. Lot Fourteen provides more than just a development blueprint during a period of profound change; it teaches communities how to strategically plan, analyze, and innovate.

References

Architecture, AU (2023) Tarrkarri, Adelaide's aboriginal art museum, hangs in the balance. https://architectureau.com/articles/adelaides-aboriginal-art-museum/

Cheshmehzangi A (2020) Reflecting on urban identities. In: Identity of cities and city of identities. Springer Singapore, Singapore, pp 133–178

Cheshmehzangi A (2025) Relations in Chongqing: interactions and connections within a multi-identity city. In: City walks in Chongqing: mapping a multi-layered, multi-identity, and multi-locality city that never sleeps. Springer Nature Singapore, Singapore, pp 159–173

Cunningham S, McCutcheon M (2020) Australian cultural and creative activity: a population and hotspot analysis-innovation precincts in Adelaide

Cunningham S, McCutcheon M (2023) Rearticulating the creative industries-STEM relationship: the case of innovation precincts in South Australia. Creat Ind J 16(1):22–41

McIntyre P, Kerrigan S, McCutcheon M (2021) Australian cultural and creative activity: a population and hotspot analysis: Marrickville. https://eprints.qut.edu.au/208593/

Woods Bagot (n.d.) Occurring naturally. https://www.woodsbagot.com/projects/aboriginal-art-and-cultures-centre/

Chapter 20
Reimagining Fishermans Bend: Recasting the Industrial City for a Mixed-Use Future in the City of Port Phillip, Melbourne

20.1 The Urban Frontier: Fishermans Bend and the Ambition of Scale

Fishermans Bend, a daring reimagining of 480 hectares of inner-urban industrial land on the outskirts of Melbourne's central core (Victoria Government n.d.), is one of the most ambitious urban changes in Australian history. The region has long been in the city's blind spot because of its expansive warehouses, haphazard street layout, and close proximity to the working ports of the Yarra River. However, Fishermans Bend is now at the forefront of a new urban vision: a mixed-use, walkable, dense, and innovative district that is expected to house 80,000 jobs and more than 80,000 residents by the middle of the century (Victoria State Government 2018). As a result, it is more than just a redevelopment; it is a rewriting of Melbourne's economic and spatial DNA, providing a chance for future generations to reconsider what it means to create urban life out of industrial land.

Fishermans Bend, which is situated immediately southwest of the central business district and is bounded by Port Melbourne, Southbank, and Docklands, is a strategically located yet historically disjointed location. It has remained predominantly car-dependent, underserved by public transportation, and characterized by monofunctional land uses despite being close to the city center. Up until recently, the urban experience inside its borders lacked the permeability, architectural identity, and civic life layering that inner Melbourne neighbourhoods are known for. Nevertheless, its size, prominence, and industrial vacuum provide the foundation for a reinvention in the twenty-first century. Fisherman's Bend is simultaneously an experiment in foresight and a restoration task.

Melbourne's planning response has sought to manage this complexity through the articulation of a coordinated vision known as the Fishermans Bend Framework. This strategy puts integrated planning at the forefront, in contrast to earlier iterations of Docklands or Southbank, which were criticized for their disjointed delivery and unimpressive public areas (Keys et al. 2024). With its own personalities, land uses,

A. Cheshmehzangi, *Australia's Urbanity*, Urban Sustainability,
https://doi.org/10.1007/978-981-95-3881-2_20

and public space policies, the five precincts established by the framework are Lorimer, Montague, Sandridge, Wirraway, and the Employment Precinct. The goal of the plan, which is to build a networked collection of urban environments that reflect various economic roles, housing demands, and cultural expressions rather than a single monolithic neighbourhood, depends on the diversity of these sub-precincts.

The idea of mixed-use intensity is essential to this change. Fishermans Bend seeks to create a complex blend—where living, working, learning, and recreation are tightly woven into the daily rhythms of urban life, instead of separating industrial, residential, and commercial purposes. This comprises adaptable spaces made to adapt to shifting social and economic circumstances, schools incorporated into walkable centers, and high-density housing situated close to employment clusters. With the help of a vast network of parks, streetscapes, public transportation, and social infrastructure, the precinct is envisioned as a location where families, students, small enterprises, and tech start-ups coexist alongside innovative manufacturing.

The Employment Precinct, which is expected to develop into a center of innovation in sophisticated manufacturing, aerospace, and defence, is especially noteworthy. This region, which is supported by partnerships with academic institutions, research centers, and important government organizations, is not only a hub for the economy but also a representation of Melbourne's transition from a post-industrial legacy to a knowledge and innovation economy. Its success hinges on the development of a thriving urban ecosystem that fosters inclusive economic participation, talent attraction, and creative collision in addition to high-tech infrastructure.

However, the magnitude of ambition entails a great deal of complication. This kind of urban revitalization necessitates patient governance, long-term civic leadership, and innovative design in addition to strong infrastructure planning. The everyday urbanity that creates a sense of place and community, e.g., cafés, cultural venues, easy navigation, and human-scaled activity, is still missing from a large portion of Fishermans Bend. The initial phases of development must work twice as hard to establish identity, engage streetscapes, and build confidence in the precinct's future. At this point, placemaking becomes not only desirable but also necessary—not as a superficial adornment, but as a fundamental tactic to infuse civic life into an area that has previously operated outside of the urban imagination for decades.

Fishermans Bend defies several Australian urban planning norms. It's not a straightforward greenfield growth or a brownfield cleanup. It is an active testing ground for ideas, examining how cities may provide infrastructure with personality, jobs with liveability, and density with dignity. It challenges the ability of a city like Melbourne, which is renowned for its suburban sprawl and gridded past, to build a new urban center that addresses social diversity, future industries, and climate realities. Can we scale without sacrificing character, it asks? Is it possible to densify without compromising humanity? Is it possible to turn a fringe area into the bustling center of city life? Fishermans Bend is still a work in progress as the design changes and construction moves along gradually. Even so, it serves as a mirror for Australian urbanism's future, which will be shaped by our ideas, connections, and designs for future generations as much as by the things we construct.

20.1.1 Frameworks for Change: Policy Intent and Strategic Direction

Fishermans Bend started with a daring—and contentious—act of rezoning rather than a cohesive vision. Without a thorough plan, the Victorian Government rezoned vast tracts of industrial land around the precinct for capital-city use in 2012 (Felson and Langenheim 2022). In addition to raising concerns about disorganized growth, inadequate infrastructure, and lost chances for urban cohesion, this bold action marked the start of an ambitious makeover. Fishermans Bend was progressively transformed into one of Australia's most carefully planned urban regeneration zones through a series of corrective and progressive planning initiatives.

A fundamental conflict between the need for intentional public policy and market-driven urban change was made clear by the early rezoning decision. While planners, urbanists, and community groups expressed concerns about the lack of social infrastructure, open space, and transit planning, developers seized land quickly because they saw an opportunity for high-return residential skyscrapers. A state-led recalibration resulted from this misalignment of scale, vision, and preparation. To spearhead the creation of a more integrated and design-led framework for the district, the Victorian Government "called in" all significant planning applications in 2015 and formed the Fishermans Bend Taskforce.

From reactive rezoning to proactive precinct-making, this signalled the start of a dramatic change in government. Together with technical experts, community stakeholders, and local municipalities, the taskforce started the meticulous process of defining what Fishermans Bend could and ought to become (City of Melbourne n.d.). The outcome was the Fishermans Bend Framework, which was formally unveiled in 2018 and offered a precise and ambitious road map to direct the change over the ensuing decades.

At its core, the framework is in keeping with Plan Melbourne's metropolitan planning goals, which establish Fishermans Bend as a key element of the city's plan for long-term, sustainable population and job growth. The framework aims to avoid the mistakes of previous renewal programs by incorporating liveability, diversity, and environmental performance principles from the beginning (Tara et al. 2018), with forecasts of 80,000 new residents and an equivalent number of employment. This covers specifications for accessible housing, walkability, integrated transportation systems, and access to green space, work, and education.

Maintaining this convergence between vision and action has been largely dependent on governance. Both the City of Melbourne and the City of Port Phillip are included in the precinct, necessitating constant collaboration between state planning organizations, infrastructure agencies, and municipal governments (Victoria State Government 2018). The government created the Fishermans Bend Development Board, a multi-agency governance body, to facilitate this. Its job is to make sure that planning controls, investment choices, and delivery methods all contribute to the long-term goal. This degree of control demonstrates an awareness that lasting leadership and strong governance are just as important for large-scale revitalization as planning policies.

Importantly, the framework is a spatial translation of objectives and values rather than merely a land-use plan. Given that the precinct is situated inside an estuarine terrain that is prone to flooding, it enshrines the importance of environmental care. With the help of new tram and bus routes as well as long-term rail provision, it requires a compact urban shape to lessen reliance on cars. It incorporates sustainability standards for buildings and precincts, such as water-sensitive urban planning, circular economy ideas, and renewable energy requirements (Brotchie 2024). Additionally, it incorporates community infrastructure and culture as essential elements of liveability, guaranteeing that recreation, health care, and education are integrated into the city from the outset.

The strategic direction's emphasis on economic identity is another characteristic (Cheshmehzangi and Munday 2022) that sets it apart. Fishermans Bend is intended to be an employment-led project, in contrast to previous rejuvenation initiatives that placed a strong emphasis on residential uplift. Adjacent to the new engineering campus of the University of Melbourne, the Employment Precinct was established to anchor the site in a knowledge economy future that is linked to sophisticated manufacturing, research, innovation, and creative sectors. In order to maintain affordability, accessibility, and industrial demands over time, this goal necessitates meticulous land use and infrastructure calibration.

The Fishermans Bend Framework's awareness that size necessitates planning is arguably its greatest accomplishment. The precinct's transformation will not be delivered overnight; it will unfold over several decades. The framework has prevented ad hoc development and given communities, designers, and investors alike certainty by establishing a solid policy spine and strategic direction early on. Thus, Fisherman's Bend serves as a reminder that ambitious urban initiatives require patience, good governance, and a common vision of the kind of city we wish to leave behind. Future generations' lives, careers, and sense of belonging will be influenced by the structures we establish today.

20.2 Designing the Future Grid: Built Form, Mobility, and Mixed-Use Integration

Fishermans Bend's scope and aspirations necessitate a flexible, multi-layered, and forward-thinking spatial logic rather than a static masterplan (City of Melbourne n.d.). A new urban grid that embraces complexity and departs from the previous patterns of industrial monoculture is at the core of this developing vision. It is a grid where urban diversity, mobility, landscape, and density come together to create flexible, mixed-use communities. The architectural concepts of the precinct aim to create a long-lasting urban framework that can accommodate many uses, scales, and rhythms of city life over the ensuing decades, rather than just focusing on instant transformation.

Each of the five precincts that make up Fishermans Bend, Lorimer, Montague, Sandridge, Wirraway, and the Employment Precinct, has a specific function within the larger scheme of things. In addition to being rigorously defined for administration, these sub-precincts allow for locally unique building forms and functions while maintaining the coherence of the entire regeneration region (Shaw 2018). A hierarchy of streets, laneways, boulevards, and public areas that enhance legibility and connectedness is supported by the purposefully fine-grained, walkable urban grid that is suggested across these precincts.

The permeability principle is essential to this grid logic. Fishermans Bend suggests a highly integrated street network, one that prioritizes cycling and pedestrians without sacrificing the requirements of public transportation or logistics, in contrast to the superblocks and wide-road infrastructure that once characterized the area. The plan's mobility corridors are characterized as dynamic urban areas intended to promote interaction, movement, and amenity rather than only serving as transportation lines. With the help of new bike superhighways and pedestrian-focused spaces, future tram lines will pass through important precincts, guaranteeing that every area of the precinct is reachable by public transportation and green space.

Green infrastructure and the public domain serve as the future grid's connecting threads. The precinct's integrated stormwater corridors (Joyce et al. 2020), linear parks, and urban plazas are all intended to serve as ecological and climate resilience measures in addition to being recreational features. While buildings are governed by restrictions that guarantee active frontages, visual permeability, and flexible usage throughout time, streets are designed to support huge canopy trees, passive watering, and layered mobility. The architectural approach purposefully blends landscape and infrastructure rather than considering them as distinct realms, making waiting for a tram or strolling down a green corridor a social and civic experience.

Another fundamental component of the built environment concept is mixed-use clustering. The framework encourages the vertical and horizontal integration of residential, commercial, and civic purposes rather than zoning sizable areas of the precinct for single uses. Important nodes like Sandridge and Montague are intended to be dense urban towns that combine street-level retail, commercial, and cultural programs with mid- and high-rise residences. The built form approach is purposefully diverse, providing a range of perimeter blocks, podium towers, and finer-grain infill that may change to meet community and market demands. Design guidelines encourage passive design, modular construction, and materials that reflect the area's industrial heritage while promoting architectural innovation.

Crucially, the framework strikes a balance between human-scaled design and long-term density objectives. The urban design is still based on the notion of neighbourhood character, even if Fishermans Bend expects substantial vertical growth. In order to improve legibility and lessen overshadowing of public space, building heights are focused along important transit spines and step down toward sensitive margins, such as the Port Melbourne interface. An urban rhythm that promotes both liveability and economic intensity is made possible by transitional typologies, which permit change in form and use. This responsiveness is reinforced by the focus on adaptable buildings, which are ones that can accommodate various uses throughout time and guarantee that the built environment may develop alongside the community.

Mobility is the vascular system of urban life in this futuristic city; it goes beyond simple transportation. By making walking and bicycling safe, easy, and pleasurable, the design approach facilitates a modal shift away from reliance on cars. However, the Employment Precinct's freight and logistics regulations recognize the continuous requirement for operational infrastructure. This balance is reflected in the street hierarchy, which keeps local streets and corridors with high traffic separate while guaranteeing smooth mode transitions.

Lastly, the logic of the grid incorporates the identification question. The goal of Fishermans Bend is to create its own urban identity by combining the industrial heritage, modern aspirations, and ecological dedication, rather than to imitate other inner-Melbourne neighbourhoods. It is a location that was intended to be a developing city district rather than a final product, with distinct spatial objectives that allow for cultural layering and evolution. Fishermans Bend represents a new sort of urbanism by creating a grid that supports complexity, movement, growth, and memory (Cheshmehzangi 2016). In this type of urbanism, identity and infrastructure reinforce each other (King 2004), and the physical environment serves as a stage for the development of urban life rather than merely a backdrop.

20.3 Innovation and Inclusion: Economic Ecosystems and Everyday Life

Redefining how we live, work, and interact with one another in a new type of urban region is the goal of Fishermans Bend's metamorphosis, which goes beyond simply recreating the land. The project's size and scope enable a deliberate blending of community involvement and economic innovation in an effort to create a city district that is as socially and economically vibrant as it is economically. The development of commonplace areas and experiences that foster social interaction, equity, and urban life is receiving just as much attention as high-value industries like advanced manufacturing, design, and technology, which are essential to the precinct's economic plan.

The concept of an innovation ecosystem—not just a collection of industries, but a dynamic network of production, co-location, and information exchange—lays the foundation for this goal. With the Employment Precinct serving as a catalyst for job development and worldwide competitiveness, Fishermans Bend has been designated to draw investment from aerospace, clean tech, defence-related R&D, and high-end design industries. However, this ecology is being incorporated into a mixed-use urban setting, where public life, housing, education, and culture all share the same physical environment, in contrast to traditional industrial zones. This strategy makes sure that social and civic realms are strengthened rather than isolated from economic progress.

Fishermans Bend's workforce and identity will be greatly influenced by the presence of innovation and education anchors, such as universities, research institutes, and industry-led training centers. These institutions are envisioned as open, accessible locations woven into the urban fabric, where residents, businesspeople, and students interact in common areas rather than as isolated campuses. Fostering informal education, intergenerational communication, and a thriving public life that encourages innovation and teamwork are the objectives.

Equally important is the precinct's commitment to social inclusion and housing diversity. While Melbourne's inner suburbs have often struggled with affordability and exclusion, Fishermans Bend is being planned with a strong mandate to deliver a mix of housing types, tenures, and price points. The strategy includes provisions for social housing, key worker accommodation, and adaptable housing models that respond to changing demographic needs. Rather than segregating affordable options, the intent is to weave them into the urban fabric—to ensure that people from varied income levels can live in proximity to jobs, services, and green space. This spatial and social integration is critical to the precinct's long-term sustainability and resilience.

In order to humanize the precinct's scope and aspirations, placemaking is essential. The public domain is being thoughtfully curated for experience as well as aesthetics and functionality, acknowledging that innovation flourishes in settings of encounter and exchange. Streets are made to be friendly and walking. Events, markets, concerts, and casual get-togethers are planned for parks and plazas. Particularly along the Yarra River, waterfront areas are positioned as hubs for cultural celebration, communal recreation, and climate-adaptive landscape architecture. It is anticipated that these facets of public life would give rise to the city's daily rhythm, giving locals, employees, and tourists a sense of place and ownership.

Cultural infrastructure is also receiving special attention. Fishermans Bend is intended to be a cultural landscape where art, legacy, and identity are evident in the public space and constructed form, in addition to being a business district. The goal of historical interpretation, public art commissions, and indigenous design concepts is to instil memory and significance in the newly created urban landscape (Hayden 1994). The precinct seeks to create a unique sense of place that speaks to its past and its goals by utilizing the site's industrial history, marine heritage, and cultural variety.

The ability of Fishermans Bend to foster human flourishing in addition to innovation and growth will ultimately determine its level of success. This entails creating for communities rather than just goods, and for inclusivity rather than exclusivity. It entails developing an urban neighbourhood where people from all backgrounds can live side by side—where a young family, a student, a café owner, and a robotics researcher can all see a bright future. The chance to demonstrate that innovation and inclusion are allies rather than enemies in creating the city of the future is as great as the challenge.

20.4 From Vision to Reality: Lessons in Long-Term Urban Transformation

One of Australia's most ambitious urban redevelopment initiatives, Fishermans Bend is a daring attempt to transform a sizable area of underutilized industrial property into a vibrant, future-ready zone. However, bringing a vision to life is rarely simple. The precinct provides a thorough and enlightening glimpse at the realities of city-making at scale as it transforms from a policy framework to a living environment. It serves as a reminder that creating the urban future calls for more than just creativity; it also calls for decades of perseverance, teamwork, and flexible tactics.

The enormous amount of time that Fishermans Bend requires is one of its most distinctive characteristics. This precinct is designed to evolve over several decades, with land parcels changing at varying rates and in reaction to market cycles, infrastructure delivery, and community requirements, in contrast to traditional projects that take five or 10 years to complete. Maintaining coherence, relevance, and momentum throughout generations of political leadership, community expectations, and economic shifts is a complex challenge brought on by this lengthy horizon. Therefore, urban development at this scale needs to be imaginative without losing sight of practical implementation, and adaptable without becoming fragmented.

Both the project's potential and complexity have been significantly influenced by governance. The state's assistance in creating a stronger planning framework was crucial to realigning the project toward public interest outcomes after it was initially tainted by speculative rezoning and disjointed oversight. Fishermans Bend is currently moving toward a more integrated future with the support of a committed authority, a long-term planning strategy, and more precise policy guidance. Nonetheless, there are numerous lessons to be learned about the significance of early governance clarity, open communication with stakeholders, and a readiness to adjust course when the first momentum falters. Building a city involves both physical infrastructure and political architecture.

The staging of infrastructure presents an additional layer of difficulty. Sequencing is crucial for a site this size; utilities, open space, schools, transportation, and digital infrastructure must all come at the appropriate time and location. Isolated enclaves could result from premature development, and public and investor confidence could be weakened by delayed investment. To fully realize Fishermans Bend's potential, public transportation integration, especially future tram links and metro extensions, will be essential (Victoria State Government 2018). This illustrates how crucial it is to integrate precinct design with the schedules for urban infrastructure in order to combine ambition with delivery capacity.

In order to maintain momentum and character over time, urban design is particularly important. Strong design concepts, including human-scale density, green networks, a variety of building typologies, and the adaptive reuse of heritage structures, have been established by Fishermans Bend's precinct frameworks and structure plan. Plans alone, however, do not create urbanity; repetition, experimentation, and lived experience do. Perhaps the ultimate test of urban design discipline in long-term regeneration initiatives is the ability to permit incremental development while preserving coherence and character.

Above all, Fishermans Bend urges us to reconsider what post-industrial areas can become in light of today's issues, such as social cohesion, housing demand, economic transition, and climate change. It emphasizes that change is about development rather than erasure, about integrating the past into the future and establishing aspirations in reality based on place. Fishermans Bend serves as both a warning and an example for other cities wishing to refit legacy urban lands: a living laboratory where the politics and art of transformation are being gradually revealed.

References

Brotchie R (2024, October) Creating a water sensitive city strategy for fishermans bend urban renewal area in Melbourne, Australia. In: Creating a water sensitive city strategy for fishermans bend urban renewal area in Melbourne, Australia. Water Environment Federation.

Cheshmehzangi A (2016) Temporary and temporality: public realm regeneration through temporary events. J Urban Regener Renew 10(1):58–72

Cheshmehzangi A, Munday R (2022) From global to local: the case of migration and urban identity for regenerative city transformations. J Urban Regener Renew 15(3):308–324

City of Melbourne (n.d.) Fishermans bend urban renewal. https://www.melbourne.vic.gov.au/fishermans-bend-urban-renewal

Felson AJ, Langenheim N (2022) Fisherman's bend, Victoria, Australia. Landsc Archit Sea Level Rise 90–102

Hayden D (1994) The power of place: claiming urban landscapes as people's history. J Urban Hist 20(4):466–485

Joyce P, Brotchie R, Howard D (2020, March) Fishermans bend water sensitive drainage and flood strategy. In: National stormwater symposium 2020. Water Environment Federation

Keys E, DeGruyter C, Pemberton S (2024) Melbourne's consolidation: integrated planning in practice. In: Australasian transport research forum 2024 Proceedings, 27–29 November, Melbourne, Australia. https://australasiantransportresearchforum.org.au/wp-content/uploads/2025/02/ATRF2024_Resubmission_25.pdf

King A (2004) Spaces of global cultures: architecture, urbanism, identity. Routledge

Shaw K (2018) Murky waters: the politics of Melbourne's waterfront regeneration projects. Urban regeneration in Australia. Routledge, New York, pp 135–158

Tara A, Ninsalam Y, Anderson C (2018) Interrogating urban renewal scenarios using skyline analysis in fishermans bend, Melbourne. J Digit Landsc Archit 3:2–11

Victoria Government (n.d.) Fishermans bend: Australia's largest urban renewal project. https://www.vic.gov.au/fishermans-bend

Victoria State Government (2018) Fishermans bend framework: the next chapter in Melbourne's growth story. https://www.portphillip.vic.gov.au/media/kwtjbwoe/fishermans-bend-framework-october-2018.pdf

Chapter 21
Macquarie Point Urban Renewal: The City Reimagined Through Renovating Underutilized Land on Hobart's Edge

21.1 Port to Precinct: The Regeneration of Macquarie Point

Once a gated industrial outpost on Hobart's waterfront, Macquarie Point is now the center of a large-scale urban renewal initiative aimed at redefining the city's relationship with innovation, the environment, and civic life (Hill 2019). The location served as a logistical port support zone and operating rail yard for many years; it was hidden from view, physically near the city center but psychologically far away. It now serves as a backdrop for one of Tasmania's most revolutionary urban initiatives, reflecting a larger change in Hobart's development story from closed off and service-focused to open, flexible, and forward-thinking.

More than merely repurposing abandoned land, Macquarie Point's regeneration aims to reestablish Hobart as a city that recognizes the relationship between site, purpose, and creativity (ABC News 2024). The site has unique geographic significance due to its strategic location at the meeting point of the River Derwent and Hobart's city center (City of Hobart 2018). It is the link between Tasmania's capital and its most recognizable natural and commercial landmark, making it more than just beachfront land. It is therefore regarded as a place where past and future uses, industrial function and civic aim, natural rhythm and urban ambition may all be reconciled.

The realization that Macquarie Point offered a unique, yet underutilized opportunity led to the decision to revitalize the area. In contrast to a city like Hobart, which is characterized by a compact grid and steep surrounding topography, the site was left as a huge expanse of flat ground when heavy rail equipment was moved, and port-related operations were scaled back. Instead of rushing into commercial redevelopment, planners and leaders recognized an opportunity to create a whole new zone that could represent Tasmania's increasing prominence in areas such as sustainable innovation, marine science, the creative industries, and cultural expression.

A. Cheshmehzangi, *Australia's Urbanity*, Urban Sustainability,
https://doi.org/10.1007/978-981-95-3881-2_21

As a result, the Macquarie Point Urban Renewal project became a crucial tool for establishing a new identity founded on cooperation, inclusivity, and ecological care as well as for realizing Hobart's potential as a hub for the knowledge economy. It was never intended to be a single-use area or just a business district that extended into the central business district. Rather, the location was positioned as a "generative precinct"—an urban catalyst that may foster new business ventures while enhancing the city's civic and cultural life. Macquarie Point provides a counter-narrative to Hobart's conventional planning inertia by physically and programmatically opening up the land (Tasmanian Planning Commission 2025), indicating a willingness to go past preservationist urbanism and toward intentional, design-led renewal.

Crucially, the redevelopment process has developed over a protracted period of engagement, strategic recalibration, and changing visions rather than being imposed from above. The project has sought to strike a balance between several goals since its inception: to respect the site's complex past, to restore ecological systems, to promote economic development, and to provide significant public space. These objectives are portrayed as complementing elements that help to define the precinct's identity rather than as conflicting demands. The project has continuously presented Macquarie Point as a new form of urban commons, whether through the incorporation of First Nations perspectives, dedication to zero-waste building, or the prioritization of public access to the waterfront.

Macquarie Point represents a significant turning point in both symbolism and reality within the larger framework of Hobart's development. The city has long been praised for its artistic vitality, heritage charm, and natural beauty, but it has also struggled with issues of climate resilience, economic diversification, and spatial inclusivity. These issues are directly addressed in this precinct (Tasmanian Planning Commission 2025). Within walking distance of the city center, it enables Hobart to test with innovative employment precinct models, mixed-use density, and environmentally friendly design. As a result, it serves as a testing ground for broader cultural changes in the city's outlook on the future as well as for planning and architectural solutions.

Another factor contributing to Macquarie Point's strategic significance is its capacity to link emerging regional and international networks with Hobart's current urban fabric. It is connected to global scientific communities through its use as a location for Antarctic research and marine innovation, and it is strongly rooted in the local economy thanks to its integration of cultural infrastructure and tourism potential. This contrast between local roots and global relevance is one of the project's strongest points. While boldly entering broader discussions about sustainable urbanization and innovation precincts, it enables Hobart to maintain its distinct identity.

Finally, Macquarie Point's restoration involves more than just renovations. Reviving Hobart's urban confidence and proving that even smaller capital cities can be audacious, astute, and forward-thinking in their approach to place-making are the goals. The site becomes a statement of what cities can become when they design for meaning, inclusion, and resilience in addition to growth as it transforms from an industrial wasteland to a thriving civic and commercial landscape. In this way,

Macquarie Point represents a new chapter in the history of urban Australia rather than merely a precinct.

21.1.1 Vision, Policy, and Place-Making Mandates

Macquarie Point's revitalization has been driven by a combination of community cooperation, government leadership, and imaginative planning rather than a single mandate (Tasmanian Government 2025). At its heart is a dedication to transforming Hobart's waterfront into a significant platform for Tasmania's civic and economic development, as well as a location for urban rejuvenation. The history of Macquarie Point illustrates a changing conversation between aspiration and location, vision and viability, and local identity and global connectivity (Cheshmehzangi 2015), spanning from early frameworks to more recent legislative changes.

The precinct's potential as a national asset was the first factor to influence its strategic orientation. It was ideally situated for targeted redevelopment because it was close to important infrastructure, surrounded by scenic views, and within walking distance of Hobart's central business district. Strong political will was indicated by the initial state and federal support, but a more cogent development trajectory was solidified with the creation of the Macquarie Point Strategic Framework. By going beyond transactional rehabilitation to promote public value and strategic results, this paradigm positioned the precinct as a location for cultural identity (Cheshmehzangi 2020), economic diversity, and long-term sustainability.

The focus on integrated land use, precinct-scale thinking, and a design-led approach that prioritized social and environmental results in addition to economic objectives were essential components of this process. The project's policy foundations specifically promoted knowledge-based jobs, creative sectors, and innovation ecosystems. The government decided to use a staggered and curated delivery strategy instead of giving land to the highest bidder, making sure that long-term adaptability and public benefit were paramount throughout the implementation process.

In tandem, there was a strong directive to incorporate First Nations inclusion and cultural expression into the precinct's very fabric. In order to combine the site's complex industrial, imperial, and cultural past with a vision for future municipal identity, the regeneration was never intended to be a clean slate. This was particularly evident in the Truth and Reconciliation Art Park's inclusion as a foundational gesture—a location that represents communication, remembrance, and progress. These goals demonstrated a stronger dedication to place-based justice and co-creation and were not merely symbolic; they were reflected in design briefs, collaboration arrangements, and programmatic objectives.

The precinct's vision and delivery strategy were greatly influenced by stakeholder participation. Community consultations were viewed as chances to reevaluate ambition through public input, rather than as box-ticking exercises, from the early master-planning stages to the most current modifications. Residents, companies, educational institutions, and cultural organizations were among the local stakeholders who helped to shape the site's practical priorities and thematic direction. The end product is a development that is incredibly in touch with Hobart's personality and has the adaptability to draw in foreign partners and cutting-edge industries.

This activation-first strategy helped to embed Macquarie Point in the public consciousness, not as a construction zone but as an evolving part of the city's social and spatial rhythm. In addition to the policy and cultural layers, investment priorities were carefully aligned with infrastructure and place-making goals. Public investment was committed to the early delivery of public realm components along with remediation and site preparation, ensuring that the precinct did not remain dormant or abstract to the public. Early introduction of green infrastructure, walking paths, and event spaces signalled the precinct's civic orientation and invited early patterns of use and familiarity.

The project's success has been its ability to balance urban quality with economic opportunities. Macquarie Point adopts an open-precinct concept that integrates research, work, culture, and community, as opposed to viewing innovation as something that should be kept concealed behind campus walls or locked facilities. Zoning flexibility, sustainability goals, and cross-sectoral collaborations all promote this ambition by allowing the precinct to adapt to new trends without becoming less cohesive.

Lastly, Macquarie Point serves as an example of how urban regeneration may be viewed as an endeavour to cultivate relationships, meaning, and long-term capacity rather than only land development. The project provides a compelling model for how other Australian towns may redesign historic sites because of its inclusive vision, policy foundation, and civic-minded approach. Motivated by the idea that great cities are constructed not only with structures but also with ideals ingrained in the landscape, policy in Macquarie Point has done more than just facilitate growth; it has created a platform for possibility.

21.2 Urban Innovation by the Sea: Infrastructure, Design, and Spatial Integration

In addition to redeveloping a desirable waterfront location, the Macquarie Point Urban Renewal project aims to change the way cities like Hobart approach the future. It is a unique example of opportunity, ambition, and place coming together. The precinct is situated on the outskirts of Hobart's city center, occupying a transitional area between the past and the future, the land and the ocean. Macquarie Point has historically played a logistical function, mainly isolated from the city's daily life, as a location that was formerly dominated by industrial operations, like as rail yards and maritime facilities. A change from back-of-house to front-facing, from mono-functional land use to a hybrid, interconnected urban fabric, and from an extractive past to a regenerative future is signified by its reinvention today.

Economic, ecological, social, and symbolic goals have all been thoughtfully orchestrated in Macquarie Point's spatial design. A dedication to mixed-use development, supported by significant infrastructural investment and flexible site planning, lies at the core of the precinct. The structure is determined by a logic of convergence rather than strict zoning: mobility corridors are made to facilitate access to public transportation, bicycles, and walking, innovation spaces merge with public realm networks, and institutional sites complement cultural venues. Because of the blurring of traditional urban lines, the precinct can serve as a creative cluster, civic anchor, and place of employment.

The precinct's infrastructure's adaptability and sequencing are essential to its success. The precinct was designed with resilience in mind, catering to both long-term scalability requirements and the local topography and coastal dynamics. Stormwater management, sewage systems, and electricity supply are examples of civil infrastructure that has been viewed as both necessary and flexible, able to support phasing without compromising operational integrity. The precinct's backbone now includes underground services and future-ready corridors, allowing for responsive urban growth as partnerships grow and demand changes.

The way the architectural approach handles the waterfront itself is among its most notable features. The shoreline is used as a public interface, a promenade, a performing area, and a connection rather than just a picturesque edge. Along the beach, there are now numerous boardwalks and pathways that connect important public locations and provide additional access points from the city. The design adapts to the climate and cultural context by incorporating responsive seating, native plants, and ample shade to make the waterfront usable and welcoming all year round. Crucially, the precinct's coastal identity is enhanced by the foreshore, which is not thought of as an isolated area but rather as an ecological and social continuum.

Macquarie Point's physical form is purposefully low- to mid-rise, respecting both the human scale and the surrounding heritage environment. By avoiding the pitfalls of overdevelopment, this strategy promotes an urban character that is more grounded and encourages residence, exploration, and engagement. Important history buildings, like the Goods Shed, have been adaptively reused to give the area depth and memory while tying the new into the old. These renovated structures, which include cooperative offices, event spaces, and hospitality options that energize the precinct day and night, are not artifacts but rather hubs of activity.

The public space network, which extends beyond individual buildings, is essential to the spatial logic of the precinct. Development plots are interspersed with landscaped green corridors that provide biodiversity zones and microclimates to balance the constructed environment. By providing areas for relaxation, recreation, and unplanned interaction, these green linkages also fulfill a social purpose. Art installations, municipal lawns, and pocket parks are positioned to encourage impromptu gatherings and cultural expression, serving as a reminder that innovation is relational as well as technological.

Sustainability and inclusivity have also been considered in the planning of mobility infrastructure. With a structure that prioritizes cyclists and pedestrians, the precinct facilitates multimodal access. Prioritizing connections to neighbouring transit lines and ferry services ensures that the location serves as a well-connected node within Hobart's urban system rather than a dead end. Additionally, the transition zones between the precinct and the surrounding city have been carefully considered, with the goal of preserving permeability and softening barriers to promote everyday use and exploration.

Most significantly, Macquarie Point's spatial arrangement is intended to change throughout time. It is a dynamic platform that can adapt to shifts in population, technology, and societal demands rather than a master plan that has been set in stone. This openness, which permits a variety of tenancies and programmatic experimentation, is ingrained in the site's physical texture and planning tools. By doing this, the precinct embraces adaptability, a defining characteristic of resilient city-making, and steers clear of the pitfalls of overdetermination.

In all, Macquarie Point is a wonderful example of how design, landscape, infrastructure, and government can work together to transform not just space but also metropolitan identity. The spatial DNA of the precinct is one of shared value, permeable boundaries, and hybrid usage rather than isolation. The project transforms a previous industrial blank into a canvas of economic, cultural, and environmental possibilities via thoughtful design and systems thinking.

21.3 Culture, Collaboration, and Civic Energy

The emphasis progressively transfers from Macquarie Point's physical makeover to the life it supports as it transitions from vision to actual reality. At this point, the real pulse of urbanity starts to show—not just in the staging of infrastructure or architectural drawings, but also in the rhythms of everyday gatherings, artistic collaboration, and public speech. The goal of Macquarie Point is to create a place that feels uniquely Tasmanian, where innovation is rooted in culture and progress is inextricably linked to community, rather than just being about buildings or economic activities.

The revitalization of Macquarie Point has always involved more than just releasing unused property (Prime Minister of Australia 2023). The goal was to restore the site to the public domain, making it accessible to the Hobart community once more and enabling it to participate actively in city life. Rich public realm activations are the current manifestation of this goal. The area has begun to come alive with events, markets, exhibitions, and outdoor performances, offering fresh levels of interaction that draw in a diverse range of locals and tourists. These are not only fleeting displays; rather, they are the first stirrings of a larger cultural approach that views place as shared, lived, and changing.

Social infrastructure is central to this activation. Macquarie Point creates integrated community hubs that combine employment, education, and cultural experiences rather than building for discrete uses. Early activation has been made possible by the adaptive reuse of heritage buildings, such the Goods Shed and the nearby rail structures, through both temporary and semi-permanent applications. These include co-working spaces, Indigenous storytelling spaces, local food pop-ups, and environmental teaching initiatives. These programs not only revitalize the area but also foster a feeling of ownership and belonging, which is crucial in a city that cherishes its unique identity and close-knit neighbourhoods.

The precinct's development through inclusive design and consultative procedures further demonstrates this dedication to local involvement. In order to ensure that Macquarie Point reflects deeper layers of cultural past, engagement with Traditional Owners and local Aboriginal organizations has influenced both the landscape design and the site's storytelling. Being technologically or economically innovative is not enough for the precinct; it also needs to be considerate of the communities it serves and the land it is situated on.

The focus on teamwork is among the most encouraging aspects of Macquarie Point's rise. The precinct is designed to be a networked, porous area of possibility rather than a closed-off campus. Intersections between local artists, social entrepreneurs, research organizations, and cultural institutions are encouraged. This co-location strategy encourages both official and informal collaboration; for example, a design studio may be next to a university satellite space, and public art displays might encourage conversation among locals, visitors, and employees. These exchanges provide the civic vitality that characterizes great cities: dynamic, human-centered settings rather than merely operational systems.

Its scope of ambition and place-sensitivity are what set Macquarie Point apart in the larger national context. Innovation precincts in many cities face the risk of becoming exclusive areas for industry or technology, cut off from their surrounding social and cultural context. By incorporating public activity into its own fabric, Macquarie Point eludes that fate. Through its open public squares, climate-resilient landscapes, cultural programming, and collaborations with neighbourhood organizations, the precinct is evolving into a hub for daily interactions, community resilience, and creative energy in addition to being a driver of economic growth.

The focus on activation will only increase as longer-term trends emerge. As part of the precinct's developing character, cultural paths, interpretative art, educational facilities, and event programming are all planned. However, the groundwork is already in place: a new kind of precinct is taking shape, one that maintains civic vitality as its central urban force, promotes cross-border collaboration, and honours local culture. By accomplishing this, Macquarie Point provides a convincing example of how cities can turn abandoned industrial areas into vibrant, welcoming, and future-ready metropolitan areas.

21.4 A New Urban Chapter for Hobart: Lessons from the Edge

More than just one development, Macquarie Point is a symptom of a larger trend in the way smaller capital cities, such as Hobart, are starting to rethink their urban identities. The precinct, which was formerly a closed-off industrial area on the outskirts of the city, is currently being developed into a strategic hub of activity that is culturally anchored, environmentally conscious, and economically future-focused. It is a project that reimagines urban redevelopment in a way that strikes a balance between scale and intimacy, local integrity and global potential, and legacy and innovation.

The way Macquarie Point redefines the urban limit is among its most significant teachings. Areas like this, which are surrounded by ports, isolated by infrastructure, or characterized by a single land use, have historically been viewed as backdrops rather than catalysts. However, Macquarie Point shows how these outlying areas can once again become central, both figuratively and geographically. The precinct is contributing to redefining the waterfront in Hobart as a linking and generative edge rather than a line of division by shifting its orientation toward the sea, integrating with municipal networks, and embracing its historic and cultural background.

This shift has necessitated new kinds of governance, cross-sector coordination, and community trust in addition to design ambition. The understanding that economic change cannot be achieved at the expense of environmental degradation or social marginalization has had a significant influence on the regeneration process. Rather, Macquarie Point is being meticulously layered to make space for a slow emergence, where community relationships help form long-term vision, early activations create momentum, and development is planned out with care rather than hurriedly. A model of change that is sensitive to the scale, values, and rhythms of a city such as Hobart is thus provided by the project.

Additionally, the initiative clarifies how smaller communities might adopt innovation precincts without sacrificing their civic character. Macquarie Point is not a business clustering or tech development template that was imported. It is being developed from the ground up with an awareness of Tasmania's distinct knowledge assets, which include Indigenous knowledge systems, the arts, education, and marine science and climate resilience. These are not only resources to be used for profit; they are the intellectual and cultural underpinnings that determine the place and the people for whom it is designed. The project's focus on public space, mixed usage, and inclusive access supports the notion that co-presence, open communication, and openness are more conducive to invention than seclusion.

Perhaps most significantly, Macquarie Point demonstrates how underutilized land and aging infrastructure can be turned into shared civic futures. This is about linking past and horizon—preserving tangible reminders of location while creating a new economic and cultural narrative, rather than replacing the old with the new. The project's success will be determined by how well it connects with the city and involves people in its changing narrative, not just by the quantity of jobs or square meters produced.

Macquarie Point ultimately challenges us to reevaluate what transformation in the Australian urban setting entails, particularly for smaller cities that are situated on the periphery of bigger national networks. It serves as a reminder that artistic expression and economic growth may complement one another, that urban planning can be both ambitious and considerate, and that civic life can breathe life into even the most neglected areas of our cities when it is fostered via inclusive, deliberate placemaking. Macquarie Point serves as both a symbol and a tactic as Hobart embarks on this new urban chapter, providing insights on how to design cities for connection, creativity, and shared urban life in addition to growth.

References

ABC News (2024) The long road to Macquarie Point becoming the preferred site for a $775 million multi-purpose stadium in Hobart. https://www.abc.net.au/news/2024-09-19/hobarts-macquarie-point-history/104361312

Cheshmehzangi A (2015) Urban identity as a global phenomenon: hybridity and contextualization of urban identities in the social environment. J Human Behav Soc Environ 25(5):391–406

Cheshmehzangi A (2020) Reflecting on urban identities. In: Identity of cities and city of identities. Springer Singapore, Singapore, pp 133–178

City of Hobart (2018) Hobart: a community vision for our island capital. https://yoursay.hobartcity.com.au/29431/widgets/203618/documents/78915

Hill RJ (2019, January) Practice learnings from master planning Tasmania's gateway ports. In: Australasian coasts and ports 2019 conference: future directions from 40 [degrees] S and beyond, Hobart, 10–13 September 2019: future directions from 40 [degrees] S and beyond, Hobart, 10–13 September 2019. Engineers Australia, Hobart, pp 576–595

Prime Minister of Australia (2023) Macquarie Point to be revitalised with government co-investments. https://www.pm.gov.au/media/macquarie-point-be-revitalised-government-co-investments

Tasmanian Government (2025) Macquarie point multipurpose stadium: enabling legislation report, Consultation draft. https://www.tas.gov.au/dpac/macquarie-point-multipurpose-stadium-consultation/documents/Macquarie-Point-Multipurpose-Stadium-Enabling-Legislation-Draft-Report.pdf

Tasmanian Planning Commission (2025) Macquarie point multipurpose stadium—Representation to Tasmanian Planning Commission, Stadia Precinct Consortium. https://www.planning.tas.gov.au/__data/assets/pdf_file/0004/808699/384-Dean-Coleman-doc-1-Submitted-08-May-2025.pdf

Part V
Urban Infill and Mixed-Use Regeneration

Projects revitalizing existing urban areas through strategic densification and heritage-conscious development:

- **Epping Renewal, Sydney**—A proposed mixed-use development supporting high-density urban living.
- **Main Drive Kew, Melbourne**—A heritage-conscious redevelopment integrating modern residential spaces.
- **The Ribbon, Sydney**—A striking urban renewal project integrating hospitality, public space, and city activation.
- **Darwin City Deal, Darwin**—A federal, state, and local government collaboration to transform the city into a tropical capital with enhanced public spaces.
- **Glebe Hill Village, Hobart**—A sustainable urban village integrating retail, housing, and green spaces.

Chapter 22
Density with Dignity: Epping Renewal and the Urban Evolution of Sydney's Middle Ring

22.1 Rising from Within: Epping and the Next Chapter of Sydney's Urban Story

Once a peaceful suburban community, Epping is poised to become a paradigm for how Australian cities can develop from within. It is at a pivotal point in Sydney's urban story. The Epping Renewal, which is situated at the intersection of housing demand, infrastructural investment (Ruming and Fingland 2021), and policy-led urban intensification, signifies a fundamental rethinking of how density and liveability can coexist in a fast-expanding metropolis.

In the past, Epping was a low-rise, green residential area in Sydney's middle ring, a system of suburbs that, while having excellent connections, resisted major expansion for most of the twentieth century. However, the way this zone is perceived has changed over the last 20 years due to demands on the urban region. The center ring has become a zone of opportunity as the core city becomes more crowded, and the peripheral suburbs get farther away. Epping is a strong contender for integrated regeneration because to its advantageous position, which is anchored by a significant train station and has recently been improved with the arrival of the Metro Northwest.

Epping's makeover revolves around its function as a transit node. Land-use modifications that promote higher-density development around stations have been paired with investments in high-capacity public transport as part of Sydney's larger commitment to a polycentric urban form. The Epping Renewal takes use of this convergence of market demand, strategic planning, and transportation infrastructure to suggest a new urban precinct that is connected horizontally and vertically.

However, the idea is more than just constructing new homes. In the framework of an established metropolis, it aims to redefine what an urban center of the twenty-first century can be. The Epping Renewal seeks to show that infill growth can be a responsible use of space and the environment, that mixed-use development can improve community life, and that density can be liveable. It provides a way to alleviate

A. Cheshmehzangi, *Australia's Urbanity*, Urban Sustainability,
https://doi.org/10.1007/978-981-95-3881-2_22

Sydney's housing need by incorporating additional urban fabric into the city's current framework rather than expanding its borders.

The intentional layering of civic, business, and residential purposes in a condensed space is what sets Epping apart. The traditional development paradigm is transformed by this mixed-use ethos from a monoculture of residential buildings into a multi-faceted precinct that is teeming with daily amenities, ground-level activities, and interconnected movement networks. Through design methods that prioritize walkability, permeability, and street-level liveliness, the objective is not merely to house people but also to create an environment where life, work, pleasure, and movement meet.

Simultaneously, the regeneration plan welcomes Epping as a site of transition, both in terms of identity and land use. Longtime locals are both excited and concerned about the area's transition from a village-like suburb to a high-density center. The success of the regeneration will rely on its ability to preserve continuity in the face of change—by preserving local identity (Cheshmehzangi 2023), improving public areas, and incorporating the community in its development. Here, placemaking, both in its physical form and its process, becomes crucial. In addition to vertical structures, the goal is to foster place attachment and social cohesiveness.

The rise of Epping also represents a larger shift in Australian planning culture, moving away from reactive rezoning and toward more intentional urban renewal. In contrast to previous suburban development, which frequently took the form of fragmented developer-led infill, the Epping Renewal lays out a more comprehensive, precinct-based strategy (McGuirk and O'Neill 2002). It conveys a desire to plan urban growth, to coordinate housing (Randolph and Freestone 2012), open space, transportation, and community infrastructure in a design-led vision that prioritizes people.

Considering Epping's position in the metropolitan hierarchy, this vision is very important. The area has the ability to ease the burden of travel, promote local employment, and provide a comprehensive urban experience within a short radius because it is a node that connects to employment centers in Parramatta, Macquarie Park, and the Sydney CBD. Its makeover reflects the goal of developing 15-min neighbourhoods, where public transportation complements everyday urban life rather than takes the place of it and the majority of daily needs are satisfied locally.

Reimagining a location is only one aspect of the Epping Renewal. It involves rethinking a model. It poses the question of how to revitalize our cities without destroying them and how to create space for new life without sacrificing the elements that gave the past its significance. Epping is an example of the kind of deliberate, multi-layered, and people-focused urbanism that will characterize the next phase of Sydney's development, from growth at the periphery to internal rejuvenation.

22.1.1 Policy, Potential, and the Urban Opportunity

Epping's change did not happen overnight. It is the result of a purposeful policy framework that aims to shift Sydney's growth from an outward to an inward consolidation trajectory. Epping was recognized early in New South Wales' metropolitan planning schemes as a crucial site for infill development, mixed-use redevelopment, and transit-led intensification due to its prime location within the city's middle ring. This subsection examines how a combination of infrastructure spending, legislative changes, and demographic changes produced the favourable circumstances for the area's revitalization.

The Greater Sydney Region Plan, which established the idea of a metropolis consisting of three cities, i.e., Eastern Harbour City, Central River City, and Western Parkland City, connected by infrastructure and built around polycentric growth, is at the center of this change. At the intersection of these cities, Epping enjoys the advantages of being close to Parramatta and Macquarie Park, which are major employment hubs, as well as being a vital gateway between Sydney's more urbanized core and its suburban north. It was a logical choice for targeted growth because of its strategic positioning.

The story of Epping's rebirth has been significantly shaped by the state government's long-term emphasis on transport infrastructure. By cutting commuting times and connecting the area to a larger urban area, the Sydney Metro Northwest's extension and the addition of Epping Station to its high-frequency, high-capacity network significantly improved accessibility. The infrastructural investment was catalytic, not just utilitarian. It required developers and planners to envision a denser urban form that would meet housing demand while coordinating with mobility and sustainability objectives.

Planning constraints were modified to encourage vertical expansion in response to this possibility. Zoning and height restrictions were changed to allow for the construction of high-rise residential towers surrounding the station precinct. With this action, the strategy clearly shifted from preserving suburban form to taking use of urban possibilities. Others viewed it as a long-overdue acknowledgement of Epping's ability to change, while others were concerned about the loss of local character. By encouraging a wider range of uses and a more complex urban experience, the rezoning was used as a tool for both intensification and diversification.

However, there was a larger goal to establish an urban setting that provided more than just housing, which went beyond the specifics of zoning. Planning policy encouraged plans that combined shops, open space, community facilities, and transit connections, favouring precinct-based development over isolated buildings. The focus on permeability, walkability, and street-level interaction turned became a yardstick for successful development. In this regard, the public sphere became the focal point of the revitalization endeavour as well as its content.

The rate and character of change were significantly influenced by community adaptation. Despite their early resistance to significant change, longtime residents started to take an active role in shaping their neighbourhood's destiny. Discussions on scale, amenity, green space, and infrastructural pressures were made possible by community consultation procedures. The renovation was framed as a negotiation—a recalibration of requirements, values, and aspirations within a changing urban fabric, rather than an imposition, even though not all issues could be addressed.

A larger issue that many Australian communities face is how to strike a balance between home supply goals and liveability results, which is reflected in Epping's policymaking. Whether vertical living can be made appealing and feasible in a suburban setting is now being tested in the regeneration region. It poses the question of whether density, not as an afterthought, but as a fundamental principle, can coexist with open space, social infrastructure, and high-quality architecture.

Epping's revitalization has become a glaring urban potential as a result of the alignment of local involvement, transportation investment, and policy ambition. It illustrates how a region may be repositioned without losing its character when state-led vision is paired with place-specific planning and infrastructural preparation. Redefining what it means to live, work, and connect in a metropolitan Sydney that is learning to develop from within is part of the ongoing cultural and spatial transition.

22.2 Integrated Urbanism: Form, Function, and Mixed-Use Logic

The ambitious goal of creating a fully integrated urban environment that accommodates density without sacrificing legibility, liveability, or human scale is the foundation of Epping's transformation. Epping's revitalization is a useful example of how form and function may be purposefully combined using a mixed-use logic that places an emphasis on connectedness, activity, and spatial coherence while Sydney struggles with the pressures of metropolitan development.

A planning approach that views Epping as a unified urban area rather than a collection of separate buildings is at the core of the renewal proposal. The precinct is being constructed with a heavy focus on vertical integration rather than mono-functional buildings. In order to accommodate several levels of usage within a single footprint, residential towers are built to sit atop podium levels that include communal, retail, and commercial spaces (Rauscher 2021). This strategy embodies a fundamental principle of integrated urbanism: intensification must weave everyday life together across both vertical and horizontal dimensions rather than just stacking people.

With broad walkways, ample setbacks, and adaptable leases that allow for active frontages that support cafés, stores, small businesses, and social meeting places, design regulations have promoted ground-floor activation throughout the area. These frontages serve as the intersection of the public and private spheres, transforming architectural form into a street-level experience. The intention is to make ground plane life feel vibrant, approachable, and captivating even when surrounded by tall skyscrapers.

Another fundamental tactic that characterizes Epping's updated structure is permeability. Rather than creating superblocks or impermeable zones of density, the plan introduces a finer-grained network of pedestrian laneways, mid-block connections, and shared zones. By connecting important anchors like the transportation center, public plazas, and nearby schools, this spatial stitching promotes walkability and spontaneity and makes the precinct easier to navigate. Additionally, it breaks down scale, making sure that the city's everyday rhythms of movement and interaction, rather than buildings, dominate the experience.

Scale and modulation are balanced in the architectural shape itself. Residential towers in the station-adjacent zones allow for a higher density, but they are tapered and divided into various masses to enhance sun access and prevent visual bulk. In order to add depth and complexity and lessen the impression of monolithic development, facades are articulated using a variety of materials and setbacks. This modulation is critical in maintaining a human scale and ensuring that the vertical city remains visually and socially legible from the ground.

The constructed environment intentionally incorporates public space. The precinct's physical structure places pocket parks, courtyards, and green corridors as key components rather than as auxiliary areas. In addition to providing breathing room, social hubs, and microclimatic benefits, these green components also serve to soften the hard surfaces and density of the urban fabric (NSW Government n.d.). Such access to high-quality open space is not a luxury in an environment where high-density living is becoming the norm; rather, it is essential for inclusion, belonging, and health.

The intentional placement of civic and community uses also shapes Epping's mixed-use strategy: educational institutions, healthcare facilities, and cultural venues are situated within walking distance of homes and workplaces, supporting the goal of a complete neighbourhood. This not only creates convenience but also a shared and visible rhythm to urban life, where workers congregate for lunch, elders enjoy the park, and children walk to school as part of a common urban choreography. It is this choreography that gives substance to the idea of urbanity—not just buildings or density, but a lived experience of co-presence and vitality.

Importantly, the renewal does not assume that all activity must happen within the precinct itself. Regional integration is made possible by the design, which uses public transportation and active transportation routes to link Epping's downtown to the surrounding suburbs and job hubs. By preventing insularity, this outward connectedness promotes the development of a polycentric urban fabric in which every node contributes value without becoming independent. Epping's revitalization serves as an example of how mixed-use urbanization may be designed to accommodate intimacy and intensity in all these ways. It demonstrates how walkability and verticality may coexist, how public space can humanize density, and how architectural ambition can be grounded in ordinary life. Form, function, and fine-grained spatial logic are making Epping a better place to live, work, and belong, in addition to making it denser.

22.3 Living with Density: Public Realm, Mobility, and Everyday Vitality

Epping's change is gauged by how residents live, move, and feel in the precinct, not only by the number of new skyscrapers or the amount of accommodation available. The effectiveness of the renewal as density increases hinges on how well the public realm accommodates daily life and how social and physical infrastructure work together to provide a space that is not only compact but also captivating. The goal of Epping is to make high-density living a lifestyle enhanced by accessibility, diversity, and vibrancy rather than a compromise.

Rethinking the public sphere is essential to achieving this goal. The regeneration approach incorporates plazas, shared roadways, linear greenways, and gathering nodes into the precinct's structural logic rather than viewing open space as leftover or dispersed. These areas are intended to serve as both practical and symbolic hubs where locals may congregate, socialize, and take part in neighbourhood civic life. For example, the station's surrounding plazas are framed by bustling ground floors and shady seating, beckoning both casual use and daily gathering. These areas serve as the precinct's social hub and do more than just adorn it.

The rhythm of urban existence in Epping is largely shaped by mobility. Walking and public transportation are given priority over private vehicles in the revitalization plan's layered mobility system, which centers its circulation network around cyclists and pedestrians. People of all ages and abilities may use key streets safely and fairly thanks to their redesigned shared zones and low-speed zones. In an otherwise overcrowded environment, a slow-city vibe is maintained by wider walkways, planted verges, street furniture, and traffic-calming techniques.

It is essential that the station be incorporated into the precinct's everyday operations. Epping's station, a transit-rich node on Sydney's growing rail network, serves as a social and physical hub in addition to being a useful transportation hub. The layout of the precinct guarantees easy, readable, and friendly transportation to and from the station. The areas surrounding the station are designed to promote lingering and chance meetings, and entrances are made extremely conspicuous and accessible. What may have been a temporary space is transformed into an active municipal forecourt by the addition of retail frontages, seating places, and cultural markers. Mobility infrastructure is thus integrated into urban life rather than being outside of it.

Epping's public realm architecture is likewise guided by the ideas of accessibility and inclusivity. The precinct is standardized with universal design characteristics, including tactile paving, accessible services, step-free access, and navigation. The redesign process has taken into account the needs of varied cultural groups, elderly folks, families with small children, and individuals with impairments. Comfort and climate resilience are addressed via shaded spaces, water-sensitive landscaping, and covered walkways, especially as extreme heatwaves and unpredictable weather patterns become increasingly common in western Sydney.

The area is dotted with green infrastructure, which serves social and ecological purposes in addition to aesthetic ones. Green roofs, planted median strips, rain gardens, and pocket parks all help to control runoff, cool urban surfaces, and create microclimates that are conducive to social interaction. Because they provide chances for recreation, relaxation, and reconnecting with nature in their immediate surroundings, these green spaces are especially essential for people who live in high-density dwellings. Urban ecology becomes ingrained in city life rather than being a peripheral or decorative aspect of daily living.

Equally important is the soft infrastructure that underpins Epping's density. To address the everyday needs of a varied and expanding population, community centers, co-working spaces, childcare facilities, neighbourhood clinics, and educational institutions are being combined. These services make it easier to live locally, get support, and create social networks by supporting density rather than restricting it. The idea of a 15-min neighbourhood, where variety and proximity improve quality of life, is strengthened by the co-location of such services within walking distance.

To create civic energy, programming and activation are employed. Art installations, local festivals, markets, and performances are all envisioned as recurrent pulses that strengthen a place's identity and sense of community rather than merely being extras (Cheshmehzangi 2020). By allowing inhabitants to see themselves mirrored in the precinct's public life, these cultural layers serve to counteract the anonymity that is frequently associated with urban living.

In sum, Epping's revitalization demonstrates how service integration, mobility design, and careful public realm policies may improve living in a densely populated area. It proves that comfort, connection, and individuality don't have to be sacrificed for a high-density urban existence. Rather, the precinct creates a form of urban vitality that is both forward-thinking and grounded by making investments in the quality and inclusivity of everyday environments—where congestion is not a burden but a shared opportunity.

22.4 Lessons in Middle-Ring Regeneration: Quality, Transition, and Future Urbanity

More than just a tale of densification, the Epping Renewal represents a larger change in the way middle-ring suburbs in Australia are redesigned in reaction to urban pressures and evolving lifestyle demands. Epping offers a number of lessons that cities looking to intensify sensibly can learn from as a testbed for infill regeneration inside an established suburban fabric. These lessons emphasize that growth should be pursued alongside, rather than at the sacrifice of, liveability, identity, and community integrity.

The significance of scale management is among the most important lessons to be learned from Epping. Building taller is not enough to transform a suburban town center into a dense mixed-use precinct. It necessitates balancing various urban forms, directing massing to honour nearby residential neighbourhoods, and making sure that amenity density equals building density. This equilibrium is accomplished in Epping by carefully balancing the form, from the highest structures grouped close to the station to the medium-rise edges that connect the new construction to the low-scale neighbourhoods that already exist. The goal is to evolve rather than overwhelm, with density moving down toward well-known suburban grain.

The necessity of social infrastructure and public space developing together with physical form is another important realization. A place is not considered urban just because of its density. Epping shows that to make confined living appealing; a multilayered, inclusive, and well-designed public environment is necessary. The incorporation of plazas, walkable streetscapes, community centers, and tiny green areas guarantees that vertical living is rooted in horizontal experience. These soft components are not merely decorative; they are the binding agent that makes it possible for many communities to coexist peacefully and to feel included in a changing environment.

The Epping situation emphasizes the necessity of long-term flexibility as well. Urban regeneration is an ongoing process rather than a finished product. Epping's masterplan includes phases of transformation, flexible land use patterns, and responsive design strategies to accommodate future shifts in housing demand, demography, or mobility trends. In light of changing work schedules, climate change, and unstable economic conditions (Cardew et al. 2006), this suppleness is essential. To stay resilient and relevant throughout time, a middle-ring regeneration precinct needs to be able to modify its social and spatial makeup.

Most significantly, Epping shows how difficult it is to create urbanity from the inside out. Infill projects have to adhere to preexisting social, environmental, and infrastructure limitations, in contrast to greenfield initiatives, which work on blank canvases. This entails actively interacting with established mobility networks, ecological factors, heritage components, and community issues in Epping. The project demonstrates that revitalization is about reintegrating the past into a more inclusive and forward-thinking urban design rather than deleting it.

Epping's revitalization also reflects a growing awareness of suburban urbanism, which holds that inner cities don't have to be the only places with easy access to jobs, transit, education, and cultural events. It questions the idea that a thriving urbanity needs to be connected to a central core, arguing that new centers can support urban life if they are planned with consideration, coherence, and ambition. The takeaway from Epping is ultimately unmistakable: middle-ring revitalization is a cultural change as much as a planning instrument. It challenges us to view suburbs as dynamic areas of potential rather than as static or secondary. Here, good architecture, civic involvement, and human-scale thinking can create dense, enjoyable, forward-thinking, and deeply embedded communities. The next chapter of Australia's urban history is being written in these areas, which are emerging subtly within our cities.

References

Cardew RV, Langdale JV, Rich DC (eds) (2006) Why cities change: urban development and economic change in Sydney (vol 10). Routledge

Cheshmehzangi A (2020) Reflecting on urban identities. In: Identity of cities and city of identities. Springer Singapore, Singapore, pp 133–178

Cheshmehzangi A (2023) Regenerating the local characteristics: the impacts on city life experience and identity of Shangri-La's old town. In: Mapping urban regeneration: city life experiences in Yunnan, China. Springer Nature Singapore, Singapore, pp 219–237

McGuirk P, O'Neill P (2002) Planning a prosperous Sydney: the challenges of planning urban development in the new urban context. Aust Geogr 33(3):301–316

NSW Government (n.d.) Sydney Metro Northwest Urban Renewal Corridor: priority growth areas and precincts. https://www.planning.nsw.gov.au/plans-for-your-area/priority-growth-areas-and-precincts/sydney-metro-northwest-urban-renewal-corridor

Randolph B, Freestone R (2012) Housing differentiation and renewal in middle-ring suburbs: the experience of Sydney, Australia. Urban Stud 49(12):2557–2575

Rauscher R (2021) Middle Ring City renewal and value capture planning–Greater Sydney's Sydenham to Bankstown Urban Renewal Corridor (SBURC). Renewing cities with value capture planning: model for achieving equitable housing, public and open spaces, and sustainable transport. Springer International Publishing, Cham, pp 91–113

Ruming KJ, Fingland S (2021) Delivering transit-oriented development in lower order centres: the case of Epping town centre, Sydney. Aust Plan 57(3–4):161–176

Chapter 23
Heritage, Habitat, and Harmony: The Regenerative Urbanism of Main Drive Kew in Melbourne

23.1 Threads of the Past, Patterns for the Future: Reimagining Main Drive Kew

In Melbourne's inner-eastern suburbs, where established character and green streets predominate the urban language, the Main Drive Kew reconstruction is a striking illustration of how modern urban architecture can make peace with a complex and occasionally challenging past (Walker Corp n.d.). The precinct's makeover is as much a social and cultural reckoning as it is an exercise in architectural and urban regeneration, as it is situated on the site of the former Kew Cottages, an institution that was formerly at the heart of Victoria's approach to disability care.

Main Drive Kew is more than just a neighbourhood. Layers of architectural legacy, community action, institutional memory, and planning controversy are all present at this location. The grounds were both remote and symbolic for decades, a haven of care, seclusion, and changing notions of dignity and inclusion. In order to address this history, the transition to a residential future necessitated a methodical and thoughtful approach that combined heritage preservation with a fresh paradigm for inner-city life. The goal was clear: to create a space that would meet the social and urban demands of a developing modern metropolis while being open, linked, and sensitive to its history.

The reconstruction of the property provided a distinctive canvas. With a substantial amount of valuable property in Kew, only 6 km from Melbourne's central business district, it offered a unique chance to construct a new neighbourhood in a well-established and coveted metropolitan setting (Hartigan et al. 2021). Its worth, however, was never merely expressed in terms of physical estate. It dealt with how to respect the existing structures and narratives. The design challenge became one of careful integration, finding architectural responses and spatial rhythms that neither replicated nor ignored the past, because some historic buildings, including heritage-listed cottages, remained on site.

A. Cheshmehzangi, *Australia's Urbanity*, Urban Sustainability,
https://doi.org/10.1007/978-981-95-3881-2_23

The idea of a layered precinct was adopted by the redevelopment plan. The proposal recognized the site's topography, vegetation, and constructed fabric as assets rather than transforming it into a blank canvas. Mature trees and important view corridors influenced the scale and orientation of the built form, while the existing street patterns guided the design of the new mobility networks. Architecture was used as a means of communication between people of different ages, typologies, and generations.

State authorities, business developers, heritage specialists, and community champions worked together to shape Main Drive Kew. Even though it was occasionally complicated, this multi-stakeholder approach supported the notion that urban renewal ought to be a collaborative civic endeavour. By using a colour scheme and material logic that spoke to the site's institutional architecture—brickwork, colonnades, and pitched roofs—and reinterpreting it through a contemporary, home perspective, the concept eschewed the conventions of generic infill. In a region that has historically been homogeneous, new housing stock was created to foster social diversity and house a variety of occupants, including families and downsizers.

Most importantly, the site was made public. The once-closed and inward-looking institutional grounds were transformed into a permeable neighbourhood that was woven into Kew's larger landscape. New kinds of connection were made possible by green areas, walking paths, and civic gestures. With a combination of new homes and adaptive reuse structures on each side, the vast central park served as a civic anchor, bringing public life into an area that was previously thought of as being enclosed and segregated.

A typical urban development conflict between preservation and progress is challenged by the revitalization of Main Drive Kew. In this case, the past serves as a foundation rather than a limitation. The precinct has a distinct identity that cannot be duplicated because of the historic buildings, which serve as both architectural artifacts and narrative touchpoints (Cheshmehzangi and Willman 2023). In the meanwhile, the precinct looks ahead rather than backward thanks to the addition of diverse housing, planted walkways, and sustainable design elements.

Additionally, this proposal makes a significant contribution to the larger planning discussion in Melbourne on the reuse of institutional land. The issue of what to do with hospitals, care institutions, and schools becomes more pressing when these locations become more redundant or decentralized. The response provided by Main Drive Kew is based on tact and purpose. It demonstrates how legacy may influence modern urban forms by fostering an environment that is respectful, intelligently designed, and serves a public purpose, rather than by immobilizing them in time.

The reconstruction does not aim to erase the site's past in this way. Instead, it creates a new chapter that weaves a new pattern for the future while honouring the strands of the past. The precinct serves as a reminder that regeneration is about more than just renewal; it's also about reconciliation, reinvention, and the collective creation of place as the neighbourhood develops and changes.

23.1.1 From Institution to Integration: Strategic Vision and Urban Intention

Land usage was never the only factor in Main Drive Kew's metamorphosis from a remote institutional enclave to a bustling residential neighbourhood. Strategic alignment between policy, history, community values (Kendig 2012), and Kew's larger urban narrative was necessary for this multifaceted project. In one of Melbourne's most historically significant areas, what emerged was not simply a physical makeover but also a rethinking of how legacy, social responsibility, and modern living could coexist.

Recognizing the enormous symbolic significance of the former Kew Cottages site was at the heart of the reconstruction. The location, which had long been a facility for individuals with intellectual impairments, had been a part of an antiquated institutional care model that was socially and physically isolated from the community. The choice to reconstruct the property offered a chance to address this legacy by establishing a neighbourhood that is open, welcoming, and focused on the future. To do this, though, called for a vision that could delicately combine a new urban role with the site's history.

The strategy framework for the site's transformation was established in large part by the Victorian Government. The reconstruction was influenced by guiding ideas that placed a premium on integration, preservation, and superior design results through specialized planning procedures and heritage overlays (Butt 2024). The directive was unambiguous: this could not turn into a typical residential infill. The new precinct must be sensitive to its surroundings, honour its past, and make a significant contribution to Kew's social and physical identity. As a result, several significant heritage structures were kept and modified, serving as an anchor for the master design with constructed forms that reflected the site's original purpose while accommodating new uses.

To direct public space, built form envelopes, access networks, and land uses, a comprehensive structural design was created. The planning for Main Drive Kew placed more emphasis on character, intelligibility, and connectedness than large-scale suburban subdivisions (Jess Wilson's website 2022), which frequently prioritize efficiency and yield. The canvas wasn't empty. Instead of creating contrast, the urban planners aimed to integrate the district into Kew's already-existing fabric. Wherever possible, the existing topography and trees were kept. Historic patterns were followed in the alignment of the streets. Additionally, permeability was incorporated into the development's logic, encouraging mobility, interaction, and daily use rather than walling it off.

Another crucial element was community consultation. Many Kew residents and advocacy groups have valid worries about the site's redevelopment, including traffic, housing typology, heritage loss, and the possible diluting of the suburb's character. The rebuilding team worked directly and iteratively with stakeholders to address these issues rather than ignoring them. While maintaining the larger strategic goal, reactions to local feeling were shaped through workshops, design forums, and advisory groups. In addition to establishing public trust, this participatory method promoted a sense of collective accountability for the potential development of the area.

The architectural language of the new development was carefully calibrated. The new buildings drew inspiration from the materiality and rhythm of the ancient cottages and institutional structures, even if they were modern in purpose and performance. Pitch roofs, articulated façades, gable forms, and brick all contributed to the visual conversation between the old and the new. Crucially, the size of the new housing was kept in check to guarantee a human-centered setting; it was dense enough to accommodate a varied community without overpowering the site's subdued dignity.

The goal of the initiative included social inclusion. In order to maintain the history of care on-site, albeit in a more modern setting of community-based living, a portion of the housing was set aside for assisted accommodation. This action, which was both symbolic and useful, supported the notion that the precinct's future ought to be linked to its core values of human dignity, caring, and belonging.

The Main Drive renovation Kew demonstrates how urban transformation may go beyond the transactional logic of growth when it is supported by a defined aim and cooperative governance. It demonstrates that integration involves more than simply geographical proximity; it also entails narrative congruence, cultural sensitivity, and intentional placemaking. As a result, the location is now an essential component of Kew—not just a solitary recollection, but a multi-layered, dynamic communal area. The path from institution to integration serves as a template for upcoming revitalization initiatives that aim to repair, enhance, and rethink urban areas in addition to filling in gaps.

23.2 Living with Legacy: Architecture, Landscape, and Spatial Character

The renovation of Main Drive Kew is a prime example of a unique design philosophy that chooses to build with and around history rather than replace it. It was difficult to integrate modern housing into a historic location in a way that seemed cohesive, respectful, and rooted in place in an environment where the past is tangible and ingrained in both memory and substance. As a result, the neighbourhood is distinguished by its spatial beauty, complex textures, and careful consideration of the terrain and past rather than by its architectural spectacle.

The idea of adaptive reuse was central to the design approach. The idea welcomed the site's old structures as symbols of continuity and identity rather than viewing them as bothersome relics. A number of the previous institutional buildings, which were distinguished by their symmetrical proportions, pitched roofs, and straightforward brick construction, were kept and converted into low-impact infill, community areas, and assisted living units. Key heritage elements were maintained during their renovation, but modern additions like glazing, insulation, and interior reprogramming gave them a new residential function (Callan 2023). This adaptive logic made sure that the development drew power from its history rather than severing it.

The precinct's spatial character was significantly shaped by the landscape. The pattern of streets and housing clusters was determined by the terrain, mature trees, and open space corridors as part of the landscape-led strategy used in the concept. This choice guaranteed the precinct's openness and informality, reminiscent of a college, while also protecting priceless ecological and aesthetic resources. Curved streets, green spines, and ample setbacks established a rhythm of openness that seems unique and strongly tied to the site's original institutional character, in contrast to grid-based projects that force form over topography.

Crucially, the precinct's new housing typologies were intended to enhance rather than replace the constructed form that already existed. The new structures took inspiration from the heritage context's material palette and scale, but they were obviously modern in terms of construction and detail. Brick continued to be the most popular material, selected for its ability to blend in with the site's preexisting architectural language as well as its strength and thermal performance. The spectrum of tones, which included softer earth tones and deep reds, produced subtle variation without detracting from the overall cohesion. In order to allow the new constructions to mirror the rhythm of the retained buildings, rooflines were carefully modified to reflect traditional forms. Gabled and pitched profiles were preferred over flat roofs.

Diversity in housing was implemented as a purposeful tactic to promote long-term adaptability and community involvement. In order to prevent housing monocultures and promote intergenerational appeal, a variety of detached homes, terraces, apartments, and supported living alternatives were incorporated throughout the area. Crucially, density was dispersed rather than concentrated; to ensure connectedness and visual permeability, multi-residential buildings were situated close to important access points and natural areas. The architecture created a visually pleasing and socially intelligible neighbourhood by avoiding sharp contrasts and instead relying on subtle massing, material, and form transitions.

The public sphere was intended to serve as a binding fabric as well as an amenity. In order to create a spatial story that promoted movement, gathering, and quiet reflection, pathways, pocket parks, and communal gardens were introduced between dwelling clusters (City of Melbourne 2012). These areas frequently reinforce a sense of continuity and time by utilizing the site's natural vegetation or by aligning with historic axes. Bluestone edging, native plant palettes, and shaded walkways are some examples of landscape design elements that evoke the site's past while still addressing the needs of a climate-sensitive city in the future.

Both the nature and the architecture incorporated sustainability. Water-sensitive urban design elements, green corridors that promote biodiversity, and passive solar orientation were all incorporated. Numerous homes used natural ventilation methods, thermal mass tactics, and high-performance glazing. Stormwater filtration was included into the roadway design, and drought-resistant and ecologically valuable planting species were chosen. These initiatives demonstrate that, when driven by long-term thought, legacy and sustainability do not have to conflict; rather, they can reinforce one another.

What appears in Main Drive Kew is a blended urban setting where time is permitted to layer, rather than being a meticulous work of preservation or a polished insertion of the new. In a municipal sense as well as a historical and aesthetic one, the precinct seems to belong. By seeing the past as a foundation rather than a limitation, allowing new life to flourish in the gaps left by memory, and reminding us that some of the most enduring places are those that remember while they renew, it has completely reinterpreted what it means to "live with legacy".

23.3 A Community in Dialogue: Public Space, Social Diversity, and Place Attachment

At Main Drive Kew, the idea of community has been integrated into the precinct's very fiber rather than being an afterthought to the architectural layout. The development of common areas, social diversity, and a respectful conversation with the past are the cornerstones of the site's transformation from institutional isolation to an inclusive neighbourhood. In this context, placemaking takes on an active role, bridging historical divides, embracing diversity, and cultivating a sense of place.

The precinct's social rhythms have been shaped in large part by public space. The goal of the design was to create a civic environment that would facilitate social contact, peaceful retreat, and casual gatherings, rather than just a collection of homes. The neighbourhood's parks and natural spaces serve as community backyards, easing the transitions between nature and the constructed environment. Rather than being overly planned or decorative, these open areas are readable, spacious, and friendly, encouraging both unplanned activity and introspective seclusion. Their informality supports everyday life in ways that feel intuitive and grounded.

Importantly, the way these areas have been designed and utilized has been influenced by the site's historical memory. Community facilities and cultural venues are now housed in former hospital buildings, allowing for continuation of use while redefining their function in a way that is more inclusive. To ensure that history is not forgotten but rather recognized as a component of the place's changing identity (Cheshmehzangi 2021), interpretive signage, art installations, and peaceful memorial areas pay tribute to the old Kew Cottages' occupants and employees. These small interventions establish the community's evolution in collective memory while also fostering a stronger feeling of awareness and empathy.

Main Drive Kew's population is remarkably diversified, including a wide range of demographics such as families, downsizers, people in need of help, and professionals looking for easy access to Melbourne's inner districts. A variety of housing options as well as the creation of common areas that serve as equalizing settings—locations where individuals from all backgrounds and phases of life can interact—have helped to promote this diversity. Playgrounds, garden spaces, walkways, and communal courtyards all work as ties that promote social contact while maintaining individual privacy.

A curating style that prioritizes subtle activation over event-driven spectacle also shapes public space. Smaller-scale projects like neighbourhood gardening clubs, art classes, and walking clubs are supported by the precinct rather than depending on large infrastructure or programs to create community life. A grassroots approach to placemaking—one that arises from lived experience rather than being imposed from above—is reflected in these activities, which are frequently led by local partners or people themselves. A genuine, modest, and lasting village culture is the result.

The idea of "place attachment," or the emotional connection people develop with their environment, is also addressed by this method. At Main Drive Kew, attachment is developed gradually via everyday use, memory-making, and intergenerational interaction rather than being produced by branding or spectacle. The peaceful dignity of the neighbourhood, the ease of connecting with others, and the common knowledge that this is a neighbourhood with a complex history are some of the reasons why residents talk about feeling "at home" in addition to the design quality of their homes. A sort of temporal openness has been established by the mingling of the old and the new, inviting individuals to contribute their chapter to the site's changing narrative.

This inclusive vision has also relied heavily on accessibility. The presence of supported housing guarantees that persons with disabilities are visible and integrated into daily community life; street crossings, lighting, and signs assure safety and intelligibility; and pathways are well-connected and intended for universal access. This is a location where diversity is ingrained in the social and physical framework, not a community created exclusively for one group of people.

Main Drive Kew serves as a reminder that curating social opportunities is just as important to excellent urban redevelopment as constructing structures. The precinct has transformed from a renovation into a lived-in, dynamic, and profoundly human neighbourhood thanks to its dedication to open public space, social variety, and remembrance. It demonstrates how spaces may be created to not only house people but also to engage them in meaningful conversation with their environment, with one other, and with the common stories that drive urban life.

23.4 Lessons in Regenerative Infill: Heritage, Continuity, and the Urban Ethic

Primary Drive One of the best illustrations of how urban renewal can value the past while creating a significant future is Kew. Its transition as a former institutional site interwoven with the Kew suburb alludes to a larger urban ethic—one based on continuity, care, and the nuanced art of restrained design. This project stands out for its modest confidence that successful regeneration comes from reinterpretation rather than erasure, in a planning landscape that is frequently consumed with speed and show.

The importance of design restraint is among the most important lessons to be learned from Main Drive Kew. Cohesion and contextual integration were given priority in the development method rather than overpowering the site with new architectural statements. The old heritage structures serve as inspiration for the newly constructed forms, which respect material language, scale, and proportion. This makes it possible for the past and present to have a smooth spatial dialogue in which historic structures serve as identity anchors rather than artifacts (Cheshmehzangi 2014). A sense of continuity and humility is further reinforced by the use of classic brick, timber details, and a landscape-led layout. This lesson, that heritage ecosystems demand not only innovation but also sensitivity, is vital in an era of rising densification.

Material memory is a concept that is equally significant. Main Drive Kew is filled with physical remnants of its former existence, including carefully chosen public art, restored walkways, and conserved buildings. These components are mnemonic as well as aesthetically pleasing. They invite contemplation while letting new meanings develop since they bear the emotional and cultural burden of the site's previous uses. By grounding it in something more profound than aesthetic appeal, this method enhances the sense of location. It teaches us that regeneration should not aim to cleanse the past, but to fold it into the future in ways that are visible, accessible, and respectful.

Additionally, stewardship is essential to the precinct's long-term success. The project has been led by a dedication to governance that goes beyond delivery schedules, starting with the early masterplanning phases and continuing with community involvement. A form of custodianship that extends beyond building is demonstrated by the upkeep of open spaces, the conservation of cultural treasures, and the encouragement of local programs. This serves as a reminder that regeneration is a generational endeavour rather than a limited one. The most successful infill projects, such as Main Drive Kew, are those that give equal consideration to legacy and result planning.

The most important example of how urban infill may be a regenerative act is Main Drive Kew. It transforms a dormant landscape into one of inclusivity, belonging, and civic worth rather than merely completing holes in the city. Instead of ignoring the past, it incorporates it into a modern urban narrative where institutional history, suburban charm, and contemporary sustainability all come together. The precinct shows tradition as an enabler of distinctive and genuine urban form, challenging the idea that it is a limitation.

As Australian cities confront the need to reuse, retrofit, and reimagine legacy lands, Main Drive Kew offers a blueprint for how to do so with ethics and empathy. It demonstrates that a break with the past is not necessary for excellent urban living. Instead, it thrives when it is founded on a thorough understanding of the location, a reverence for local stories, and a sustained dedication to creating areas that are not just liveable but also meaningful. When done properly, infill holds the regenerative promise of not only filling voids but also revitalizing them with continuity and care at their core.

References

Butt A (2024) Sustainable, resilient, regenerative? the potential of Melbourne's peri-urban region. Front Sust Cities 6:1391712

Callan P (2023) The new modernist house: mid-century homes renewed for contemporary living. Thames & Hudson Australia, Cremorne

Cheshmehzangi A (2014) Spatial syntagma and identity of a place: sensing, relating to, and knowing a place. J Human Behav Soc Environ 24(7):799–810

Cheshmehzangi A (2021) Urban memory in city transitions: the significance of place in mind. Springer, Singapore

Cheshmehzangi A, Willman E (2023) The city as palimpsest to redefine the role of architectural continuity in re-establishing urban identity in post-industrial landscapes. J Urban Regener Renew 16(3):340–357

City of Melbourne (2012) Urban forest strategy: making a great city greener 2012–2032. https://www.preventionweb.net/files/36308_urbanforeststrategy2.pdf

Hartigan M, Fitzsimons J, Grenfell M, Kent T (2021) Developing a metropolitan-wide urban forest strategy for a large, expanding and densifying capital city: lessons from Melbourne, Australia. Land 10(8):809

Jess Wilson's website (2022) Preserving the character and liveability of Kew and Hawthorn. https://www.jesswilsonkew.com.au/article/2022-10-27-preserving-the-character-and-liveability-of-kew-and-hawthorn

Kendig LH (2012) A guide to planning for community character. Island Press, Washington, DC

Walker Corp (n.d.) Built with a natural sense of community. https://www.walkercorp.com.au/projects/main-drive-kew

Chapter 24
The Ribbon Effect: Designing Urban Icons for Everyday Life in Darling Harbour, Sydney

24.1 Sculpting the City's Edge: The Ribbon as Urban Catalyst

The Ribbon, an ambitious and aesthetically striking urban renewal project at Sydney's Darling Harbour, is presented in this chapter. The Ribbon is a sophisticated mixed-use project that transforms the city's waterfront edge by fusing public space, entertainment, hospitality, and urban connectedness (Guides 2019). It is more than just an architectural statement. In addition to engaging an important site, the project provides a compelling case study of how infill regeneration may be both iconic and ingrained, contributing to the changing pulse of Sydney's municipal identity.

At the edge of Sydney's iconic Darling Harbour stands a structure that has come to symbolise the city's new architectural audacity and urban reinvention, The Ribbon. The Ribbon is more than just a structure; it is an urban catalyst that brings vitality, activity, and connectivity to a previously uninhabited section of the waterfront. It represents a turning point in Sydney's changing connection with its own waterfront, where the goal of revitalization is now to re-knit the social and spatial fabric that binds the city to its waterfront as well as to increase economic return and tourist attractiveness. This section examines how The Ribbon redefines the city's edge as a site of convergence—between spectacle and daily life, between history and the future—by operating at the nexus of strong design and civic aim.

The Ribbon, which emerged from the location of the former IMAX theatre, is part of a larger story about the development of Darling Harbour (Kleeman 2019). Long considered a key civic and tourist precinct in central Sydney, the harbour has undergone waves of development since the 1980s. However, this specific section, which is situated between the freeway and the waterfront, continued to be a hole in the urban rhythm of the city for many years. Despite its utilitarian importance, the IMAX building disrupted pedestrian traffic and provided little opportunity for street interaction (Rolfe 2024). A new typology that is colossal yet activated, sculptural yet

A. Cheshmehzangi, *Australia's Urbanity*, Urban Sustainability,
https://doi.org/10.1007/978-981-95-3881-2_24

permeable, was offered by the chance to transform the site into a landmark building, addressing both symbolic and spatial disconnects.

The Ribbon's design, which is fluid, continuous, and unabashedly contemporary, responds to the precinct's experience needs in addition to its physical environment (Redpath Partners 2025). Its glass curves mirror the flow of the harbor itself, and its sinuous form rises from the ground, opening at the base to create an urban canopy that welcomes the public in. Beyond just being aesthetically pleasing, this architectural gesture represents a new vision for Sydney's waterfront, one in which the built environment interacts with the water, pedestrians, and the collective imagination of the city.

Importantly, The Ribbon is not a standalone entity. It was intended as a component of a larger public–private partnership to redefine the function of Darling Harbour in a city of the twenty-first century (NSW Government 2016). The Ribbon establishes its significance within the changing precinct by connecting rather than competing with the International Convention Centre, the redesigned public realm, and an expanding array of mixed-use developments. It turns into a spatial hinge connecting public waterfront access and urban mobility corridors, as well as the calm, immersive experience of leisure, culture, and gathering along the harbor's edge with the fast-paced flows of the central business district.

However, The Ribbon's ability to host and engage all facets of civic life is what truly distinguishes it as an urban catalyst. With a large hotel, dining and entertainment options, and new public facilities, the expansion broadens the precinct's use beyond daytime visitors or activity spikes brought on by events. With commonplace experiences integrated into the architectural scheme, it promotes a more constant and inclusive urban pulse. Instead of only consuming, the public can navigate, stay, interact, and occupy. By doing this, The Ribbon reclaims a long-peripheral, socially and geographically marginalized area of the city.

This gesture has a significant impact. For many years, infrastructure, higher roadways, restricted access, and a focus on visual branding over everyday legibility, had moulded Darling Harbour's western edge more so than urban form. The Ribbon aids in reversing this legacy with its ground-level porosity and sculptured mass (Richards 2012). Its existence creates a new connection between the horizontal and the vertical, the laneway and the landmark, the trip and the destination. It becomes a part of the city's threshold language, which allows one to go from the rigid urban core to the open, flowing harbor environment with ease.

The Ribbon's involvement in Sydney's larger urban discourse is as significant. It is a bold move in the direction of fusing civic duty with architectural spectacle. This project bases its goals on urban performance, i.e., how it facilitates pedestrian movements, creates public space, and enhances the value of both location and property in a setting where iconicity frequently runs the danger of alienating people. The Ribbon serves as a reminder that large-scale architecture may nevertheless be personal, approachable, and transformative for society.

The Ribbon is a striking illustration of how smart design may awaken hidden possibilities in underutilized locations as the city expands and reworks its post-industrial and coastal surroundings. It challenges us to rethink the edge as an interface rather than a boundary, where the city and the water connect and where public life can flourish with grace, generosity, and vibrancy. In this way, The Ribbon revitalizes the urban pulse at the harbor's edge rather than just beautifying it.

24.1.1 *From Obsolescence to Opportunity: Renewal at Darling Harbour*

Darling Harbour has long been one of Sydney's most prominent civic and entertainment precincts, but not all parts of it evolved at the same pace. Prior to The Ribbon's emergence, the western side of the harbor, which was dominated by the now-demolished IMAX theatre, served as a symbolic boundary as well as a break in space. Despite being one of the biggest movie screens in the world, the IMAX was a siloed edifice that was visually dominant, physically massive, and mostly cut off from daily life and pedestrian traffic in the city. Its existence supported a monofunctional programming approach that was out of step with Sydney's urban goals and the precinct's developing potential. The result was a daring reinterpretation of a crucial urban edge that needed to be reintroduced into the city's vitality, not just a structure.

In the mid-2010s, there were larger goals to reposition Sydney's tourism and events sector, which led to the push for rejuvenation at Darling Harbour (NSW Government 2021). To maintain the city's competitiveness on the international scene, government initiatives concentrated on developing new civic, cultural, and commercial infrastructure. This includes the International Convention Center's renovation, the remodelling of the public areas around it, and an effort to revitalize the harbor's failing areas. Due to its proximity to important transportation hubs and its role as a physical and mental barrier to improved public interaction with the waterfront, the IMAX site was recognized as one of the most strategically advantageous locations.

This setting gave rise to the notion that the location could fulfill a different function, one that involved developing an architectural form that expressed dynamism, permeability, and connection rather than merely substituting entertainment with entertainment. The goal was to turn a location with restricted access into one that was constantly bustling. The NSW Government started formulating a vision that would combine upscale leisure and hospitality with significant civic engagement in collaboration with business development companies (NSW Government 2025). The Ribbon was intended to become a distinctive building, but it had a clear municipal purpose: to open up underutilized waterfront margins, to connect the public realm across broken corridors, and to establish a new symbol of Sydney's urban design maturity.

The planning and development approval process was complex. There was pressure on one side to develop a program that would be profitable and able to support expensive real estate, such as a hotel and entertainment venues. However, the site's legacy and prominence meant that the design needed to be much more than just a showpiece. It had to demonstrate its ability to integrate public spaces, be accessible to pedestrians, and have clear architecture. This vision was refined with the assistance of important collaborations between the developer, architects, and planning authorities, transforming a previously divided land into a tiered precinct node. The formal language of the design—curving, tapering, continuous—was a reaction to the harbor's visual dominance as well as the requirement to lessen the boundaries between the street and the building.

The Ribbon was unique not only because of its size but also because of its ability to reintegrate a remote location into the social and geographical structures of the city. The project adopted an urban stitching concept, filling the gap caused by the Western Distributor, providing new views of the water, and connecting pedestrian pathways from the city center to Barangaroo. With raised terraces, benches, and covered spaces, it also pledged to create a more expansive public space that encourages unplanned use in addition to scheduled activities. A formerly contained structure was transformed into a platform for openness in this way.

Finally, Darling Harbour's transformation from obsolescence to opportunity shows how underutilized areas in desirable places can be transformed with vision, teamwork, and creative ambition. The Ribbon serves as an example of what may happen when civic objectives are given equal weight with business needs; when architecture is viewed as an invitation rather than a barrier; and when urban redevelopment is about more than just revitalization, it is about reconnection.

24.2 Architecture in Motion: Form, Function, and Mixed-Use Complexity

In addition to its sculptural presence, The Ribbon is one of Sydney's most remarkable architectural interventions in recent memory because it seamlessly blends form and function into a cohesive urban gesture. The Ribbon's architecture seems to flow between earth and sky, flowing back into the city and extending out toward the lake, in contrast to traditional towers or horizontal podiums. The dynamic nature of the city, the harbor, and the people who interact with it on a daily basis are all reflected in this building in motion, where movement is both a physical attribute and a symbolic motif.

The Ribbon's mixed-use intricacy lies at the core of its architectural philosophy. The building's commercial spaces, event and entertainment spaces, and opulent hotel are all incorporated into one continuous volume. A difficult design problem was created by this programmatic density: how to preserve architectural coherence while guaranteeing spatial clarity and user usefulness across many domains? Developing a sinuous shape that permitted both vertical and horizontal division without sacrificing overall coherence was the answer. Despite the stacking and weaving of various purposes, the building's curved façade preserves a smooth outward expression that gives it a unique and refined appearance.

From an urban perspective, the form adapts to its surroundings with sensitivity. The structure serves as a landmark and a link between Darling Harbour and the tall Western Distributor. Its curves are not random movements; rather, they are calculated reactions to limitations, protecting important view corridors, improving visual permeability throughout the site, and accommodating existing infrastructure. By navigating a difficult topography, the Ribbon's lowest levels provide new pathways and eye-catching openings that direct pedestrian traffic from the city center to the waterfront. In addition to being aesthetically pleasing, the impact has deep practical significance, influencing how people move through and engage with the surrounding precinct.

The structure breaks with the convention of venues facing inward at street level. Rather, it adopts a porous ground plane, bringing the edge to life with abundant windows, pedestrian entrances, and a vibrant public space. This is most noticeable in the terraced terraces and forecourt that make up the building's public base. These places serve as gathering places that host unofficial events, offer chances for passive relaxation, and improve the cadence of everyday life along the shoreline. They are more than just transitory areas. What may have been an elite development is turned into an inclusive public asset by The Ribbon, which breaks down boundaries and creates a variety of spatial scales.

Another important aspect of this urban choreography is materiality. Depending on the weather and time of day, the metal and glass panels covering the building's façade change in colour and reflectance. These materials were picked for their ability to blend in with the harbor environment's fluidity as well as their longevity and ease of upkeep. The project was viewed by the design team as a surface that was constantly changing in response to light, motion, and use. The Ribbon converses with the waterfront in this way rather than imposing itself on it.

On the inside, space is arranged to accommodate both the remarkable and the commonplace. While event areas are integrated into the structure to provide effective service access and seclusion, the hotel component gives vistas that frame the city and sea with equal drama. The building's circulation routes are clear and intended to guide guests through its levels without any confusion or disruption. Although there is a distinct hierarchy of areas, too much formality is purposefully avoided. The goal was intelligibility, which is sometimes overlooked in famous developments, rather than just luxury or spectacle.

The Ribbon's reluctance to compromise between style and content is what makes it such a successful architectural design. It is sculpture-like but incredibly sensitive to urban rhythms, and it is both aesthetically ambitious and functionally grounded. It skilfully negotiates the challenges of mixed-use programming, offering a model for how structures can be both remarkably shaped and functionally efficient. Crucially, it shows that outstanding design does not have to be limited to civic monuments or cultural institutions; it can and ought to be applied to business architecture that immediately interacts with the public space.

The Ribbon expands on that story by providing a fresh perspective on Sydney's edge in a city that has been defined by its relationship to water. It demonstrates how movement, material, and purpose can be combined to create an architecture that animates space rather than merely occupying it. Not just in steel and glass, but also in the daily activity that passes through and surrounds it, this is urbanity in motion.

24.3 Public Realm as Pulse: Activation, Access, and Civic Engagement

The Ribbon is a civic catalyst that revitalizes one of Sydney's most vibrant yet historically underutilized waterfront edges. It is more than just a sculpture. Its ability to excite the public sphere at ground level is its most potent contribution, even though its form may draw attention from a distance. The Ribbon's design ambition is firmly anchored in public interaction, promoting a revitalized relationship between people, place, and harbor through everything from terraces, cultural institutions, and adaptable event spaces to revitalized promenades and pedestrian access points.

The idea that urban experience is constructed from the ground up lies at the heart of the concept. Once a disjointed and abandoned urban corridor, the Ribbon's merger with the western border of Darling Harbour has made it a smooth extension of the city's cultural spine. A continuous, readable, and accessible public domain that permits people to move, congregate, stay, and engage with the harbourfront environment is essential to this change. This is about building an active civic stage where everyday life and planned events intersect, not only about transit or passive green space.

The boundaries between public space and commercial industry are purposefully loosened in the design. The base of the building is porous and aesthetically open, with numerous entry points, dynamic edges, and clear windows that engage pedestrians. Arrival is now characterized by a generosity of space that beckons, rather than by vertical separation or isolation. In addition to enhancing physical accessibility, broad promenades, terraced terraces, and raised walkways allow for a variety of social interactions. This multi-layered public space accommodates tourists, lunchtime get-togethers, morning joggers, and cultural observers.

The public realm's event infrastructure is a noteworthy aspect that has been thoughtfully incorporated to facilitate adaptable programming. The area has been planned to accommodate activation all year round, ranging from small-scale cultural pop-ups and musical performances to major civic gatherings. The lines between official venues and casual public life are blurred by these experiences, which extend beyond the structure and into the terraces, pathways, and water's edge. Temporary uses are possible without compromising daily functionality thanks to the spatial arrangement, which makes activation feel natural rather than forced.

Another important factor that promotes civic engagement is visual permeability. The building never ignores the harbor, even with its size and imposing appearance. Instead, lines of sight from the city to the water are preserved because to The Ribbon's curving form and translucent foundation. The design honours vistas' function as connecting pieces that unite the city's natural landscape and urban identity, rather than as static views. By directing people toward the river and promoting mobility across the precinct, these sightlines turn into visual invites.

Strategic connections to nearby precincts strengthen this relationship. The pedestrian flows in Darling Harbour, Pyrmont, and the central business district have been meticulously planned and incorporated into the larger circulation system. By bringing disparate areas of the city together into a more unified whole, The Ribbon transforms from a destination into a conduit. Its success is determined by the daily rhythms of urban life that flow through and around it, not only by occupancy rates or architectural praise.

Crucially, the research questions the notion that public gain must always be constrained by private development. Conversely, The Ribbon shows how a commercial project's DNA can incorporate civic giving. The development has produced a rich urban environment that enhances both the architectural skyline and the ground-level living experience thanks to cooperative planning, a thorough design review, and a dedication to place-making.

The Ribbon restores the urban heart to the waterfront in a city that has long been identified by its harbor but is frequently geographically separated from it. Moving around the city is framed as a cultural experience where access, activation, and aesthetic pleasure come together. The Ribbon sets a new standard for how modern architecture may support the civic life of its city by making public space an integral part of the project rather than an afterthought. It serves as a daring and timeless reminder that genuine urban life is lived rather than merely constructed.

24.4 Lessons in Iconic Urbanism: Integration, Intensity, and Identity

An excellent example of how iconic architecture may be firmly rooted in the realities of daily urban life is The Ribbon. It is an integrated urban precinct that successfully blends the texture, pace, and rhythm of city life with the goal of large-scale revitalization. It is more than just a single sculpture or visual landmark. By doing this, it provides crucial guidance for communities that are struggling to incorporate daring architectural gestures into preexisting metropolitan settings without devaluing the public or alienating local experience.

The significance of integration—not only at a physical level but also across functional, experiential, and civic dimensions—is among the most important lessons to be learned from The Ribbon. Because of its location on the western side of Darling Harbour, it was surrounded by a multi-layered urban setting that was already well-developed in terms of transportation, retail, tourism, and cultural facilities. The Ribbon aimed to connect and amplify these levels, becoming a link in the larger city system, rather than trying to control or isolate itself. The building provides a substantial program of hotel accommodations, entertainment, and event space, but it also maintains its porousness and connectivity to the everyday public life surrounding it. This is the key to its success.

In situations where tourism infrastructure frequently results in exclusion or oversaturation, striking this equilibrium is particularly challenging. The Ribbon adopts an alternative strategy, presenting itself as a public resource rather than a private haven. By making sure that its public space is connected with the city's and that its design language speaks to local materiality and movement, it avoids the danger of turning into an island of show. In addition to being aesthetically distinctive, the building's curved geometry gives it a feeling of softness and movement, enabling it to blend in with the urban landscape rather than stand out from it.

Furthermore, The Ribbon shows that intensity need not equate to clogged roads or isolation. To ensure legibility and usability even during busy hours, the precinct was designed for layered use, where pedestrian routes, event zones, viewing platforms, and commercial entrances coexist in spatial harmony. This multi-layered strategy promotes overlapping patterns of activity, such as visitors and office workers interacting, event attendees and locals crossing paths, and morning joggers and casual strollers sharing space. A living city is defined by this coexistence.

The importance of securing one's individuality is yet another fundamental lesson. The Ribbon's shape and goals are undoubtedly contemporary, but it finds significance in its harbor location. Its design is influenced by views, breezes, light, and movement, all of which combine to provide a space that is distinctly Sydney-specific. The building's iconic identity is emergent and site-responsive rather than abstract or imported because to this link to the location (Cheshmehzangi 2020). For global cities hoping to create destination areas without sacrificing the integrity of their urban narrative, this lesson is especially important.

Lastly, The Ribbon serves as a reminder that creating a city on this scale requires more than just a visual gesture. It has to be a charitable act of the public. Through enhanced access, activated frontages, cultural programming, and permeability, the project's contribution to the public realm guarantees that it benefits the larger community just as much as its tenants and users. It establishes a standard for upcoming initiatives in crowded cities, demonstrating that aspiration and inclusivity don't have to conflict.

The Ribbon demonstrates how to produce something spectacular while staying rooted in the everyday experiences of urban living in a world where architectural statements are becoming more and more extravagant. It teaches us how to do iconic urbanism well—carefully, clearly, and with a great deal of respect for the city it aims to create.

References

Cheshmehzangi A (2020) Reflecting on urban identities. In: Identity of cities and city of identities. Springer Singapore, Singapore, pp 133–178

Guides I (2019) Insight guides explore Sydney (Travel Guide eBook). Apa Publications (UK) Limited

Kleeman G (2019) Sydney's darling harbour: a case study in the dynamics driving change in a large city of the developed world. Geogr Bull 51(2):8–12

NSW Government (2016) The cultural ribbon draft strategy—Sydney harbour's nature and culture walk (with proposed amendments). https://meetings.cityofsydney.nsw.gov.au/Data/CulturalandCommunityCommittee/201612051402/Agenda/161205_CCC_ITEM04_ATTACHMENTA1.pdf

NSW Government (2021) Harbourside redevelopment. https://www.ipcn.nsw.gov.au/sites/default/files/pac/projects/2021/04/harbourside-shopping-centre--darling-harbour/referral-from-the-department-of-planning-industry-and-environment/dpie-final-assessment-report.pdf

NSW Government (2025) NSW means business as Australia's no.1 business events destination. https://www.nsw.gov.au/departments-and-agencies/dciths/ministerial-media-releases/nsw-means-business-as-australias-no1-business-events-destination

Redpath Partners (2025) Exploring excellence at the Ribbon Project, Sydney! https://www.redpathpartners.com/exploring-excellence-at-the-ribbon-project-sydney/

Richards DM (2012) Destiny in Sydney: an epic novel of convicts, aborigines, and Chinese embroiled in the birth of Sydney, Australia (Vol 1). Aries Books

Rolfe D (2024) Construction: solving a knotty problem. Aust Restruct Insolve Turnaround Assoc J 36(1):20–25

Chapter 25
Designing the Tropical Capital Through Climate-Conscious Urbanism: Urban Renewal and the Darwin City Deal

25.1 Setting the Vision: Darwin's Regeneration as a, Civic and Climatic Imperative

A vision that is both civically determined and climate mindful is reshaping Darwin's urban future. The Darwin City Deal represents a strategic commitment to reposition the city as a resilient, inclusive, and thriving tropical capital. It was initiated as a cooperative agreement between the federal, territorial, and local governments (Oakley 2011). This urban revitalization project is based on the facts of Darwin's distinct geographic location, changing economic foundation, and multi-layered cultural identity rather than starting from scratch. Along with financial investment, the City Deal also includes a political and geographical strategy (City of Darwin, n.d.) that views urban renewal as a means of achieving long-term change.

Darwin has always been different from other capitals in Australia. It has a history influenced by its Indigenous roots, colonial intersections, battle scars, and cyclone resilience. It is also geographically inaccessible, has a harsh climate, and is culturally rich. However, it has also long faced the problem of maintaining urban cohesion. Darwin has frequently faced a conflict between local underdevelopment and national significance due to its relatively low-density center, uneven public space, and economy that is mostly dependent on mining, defence, and federal employment (Carson et al. 2010). By making investments in the city's public life, physical identity, and infrastructure renewal—connecting urban aspiration with federal recognition—the Darwin City Deal aims to resolve this conflict.

An urban logic that prioritizes placemaking just as highly as production lies at the core of the City Deal (Hu 2019). Darwin's approach is clearly municipal in nature, in contrast to other city projects in Australia that have concentrated on significant employment centers or transportation corridors. People, location, and culture are at the heart of the regeneration agenda, which aims to create a liveable, walkable, and vibrant city that embraces diversity, adapts to the climate, and improves public life. Reimagining the urban experience in northern Australia requires building inward

A. Cheshmehzangi, *Australia's Urbanity*, Urban Sustainability,
https://doi.org/10.1007/978-981-95-3881-2_25

rather than outward, using public space, cultural infrastructure, and infill development as accelerators.

The city's tropical climate is a design need as well as an environmental limitation. In Darwin, urban discomfort is frequently seasonal yet has a significant impact on how people use space. By making investments in streetscapes, shade, water-sensitive infrastructure, and cooling public areas, the City Deal directly addresses this issue (City of Darwin, n.d.). In a tropical setting, urban comfort is not an extravagance; rather, it is necessary for fairness, inclusivity, and civic engagement. Thus, the regeneration combines urban form and climatic resilience in a way that few capital cities have ventured to do on such a large scale.

Cultural identity is also very important. Darwin's long-standing multicultural population, which includes a sizable Southeast Asian ancestry, and its strong Larrakia Nation presence make it a hub for a variety of communities. This variety is reflected in the City Deal's placemaking methods, which incorporate Indigenous design, heritage recognition, and cultural expression in the built environment in addition to consultation (City of Darwin, n.d.). The goal of the regeneration approach is to unearth and improve the tales that are already ingrained in the city's streets, buildings, and ground level rather than merely projecting a new urban image.

Another motivator is economic regeneration. Darwin is positioned by The City Deal as a gateway to the Asia–Pacific region, a city that can take advantage of its geopolitical significance and closeness to global markets. However, this goal needs to be based on a city that its citizens can live in. The redevelopment attempts to rebalance Darwin's urban economy to support knowledge, innovation, and creative industries—beyond its traditional resource base—through mixed-use infill, better mobility options, and civic anchors like the Charles Darwin University city campus. The goal of this economic and symbolic repositioning of Darwin is to make the city a desirable place to live, learn, and belong in addition to being an attractive place to invest.

The City Deal's significance is further reinforced by its tripartite structure. Seldom do local, state/territory, and federal governments have such a clear vision, budget, and delivery alignment. The institutional framework required for Darwin to achieve long-term change is provided by this multi-level governance architecture. It also draws attention to the larger national interest in making sure that Australia's urban development does not leave smaller capital cities like Darwin behind. Therefore, the City Deal is more than just a local investment (City of Darwin, n.d.); it is a sign that Darwin is important to the country's urban strategy.

Lastly, Darwin's revitalization is as much about changing attitudes as it is about construction projects. The City Deal presents a fresh story of tropical urbanization, design-led development, and cultural resilience, challenging the marginal position frequently connected to northern cities (City of Darwin, n.d.). It makes the case that Darwin's strength is in growing more uniquely itself, not that it will become a second Brisbane or Perth. The city's edge condition is being reclaimed through this revitalization, not as a frontier but as a forward-thinking metropolis where civic life, culture, and climate all come together.

25.1.1 Strategic Alignment and Shared Governance

In addition to its ambition, the Darwin City Deal is notable in the Australian urban scene for the way it was planned and implemented using a genuinely shared governance model. The collaborative infrastructure that supports the Deal is broken down in this subsection, with a focus on how federal, territorial, and local governments' alignment has made it possible to take a cohesive approach to urban liveability, economic diversification, climate resilience, and innovation. In this instance, place-specific planning and cross-jurisdictional governance (Novalia et al. 2024) are coming together to create a foundation that is both political and spatial.

The Darwin City Deal resulted from an expressly tripartite agreement, in contrast to many other Australian urban developments that are fragmented within specific jurisdictions or depend on piecemeal finance. This agreement established a clear framework for performance and a common vision among the City of Darwin, the Northern Territory Government, and the Australian Government. Fundamentally, this paradigm provided a strategic reset by shifting from reactive or transactional urban investments to long-term, coordinated place-making with reciprocal accountability and agreed-upon goals. A new type of urban intervention based on shared risk and gain was made possible by the political resolve to work together.

The City Deal's governance structure demonstrates a practical comprehension of the intricacy of modern city-making. All the levers required to address climate adaptation, draw in new companies, and provide first-rate public infrastructure are not held by one level of government. The Deal was able to overcome the typical fragmentation that frequently thwarts regeneration initiatives by fusing the territory's planning authority, the local council's local knowledge, and the federal government's finance capacity and strategic investment mandate. Crucially, this strategic alignment was not only theoretical; rather, it was enshrined in a ten-year plan that included quantifiable objectives and investment pipelines.

One of the main focuses of the Deal's governance approach has been climate adaption. All three tiers of government acknowledged that Darwin's natural setting, in particular its tropical temperature, exposure to heat, and susceptibility to extreme weather events, would influence the city's future. A proactive urban response that incorporated water-sensitive urban design, green networks, and cooling infrastructure into the regeneration process was made possible by strategic alignment around this reality (Kavaarpuo et al. 2022). Climate risk was regarded as a fundamental urban design and governance challenge that needed cross-agency solutions and finance, rather than as a distinct problem.

On the other hand, liveability was a common indicator of success. It takes careful planning to provide urban liveliness without overdevelopment in a tiny capital city like Darwin (City of Darwin 2024). A liveability-led approach was made possible by the City Deal model, in which public parks, shaded streets, and civic landmarks were not viewed as extravagances but rather as necessary to draw in new inhabitants, retain skilled workers, and promote economic growth (City of Darwin, n.d.). The agreement made sure that funds for infrastructure were allocated to initiatives that improved urban living both qualitatively and functionally.

An additional layer of strategic alignment was introduced by innovation and economic diversification. Darwin has historically relied on jobs in the public sector, mining, and defence. Governments were able to commit to a more varied economic trajectory, one that encompassed digital infrastructure, creative industries, tourism, and education, thanks to the City Deal framework (City of Darwin, n.d.). One such project that resulted from this common goal is the construction of a new Charles Darwin University city campus, which serves as further evidence that infrastructure for the knowledge economy is just as important as housing or transportation. This type of reorientation was made possible by the governance model of the Deal, which established a platform for integrated funding and decision-making and decreased intergovernmental friction.

Most significantly, the governing structure of the City Deal made it possible to be flexible and quick to react. Darwin's context—demographic, climatic, and economic—is changing quickly, and urban regeneration is not static. By allowing for gradual recalibration, the tripartite approach made it possible to handle new objectives like housing affordability or internet connectivity without encountering bureaucratic roadblocks. Large-scale urban projects rarely exhibit this kind of flexibility, which is indicative of the maturity of the intergovernmental cooperation that has propelled the Deal.

In sum, the Darwin City Deal offers a visionary yet realistic example of strategic alignment. It serves as an example of how shared governance may break down institutional silos, coordinate long-term policy goals, and open new avenues for urban investment. It demonstrates that under Australia's federal system, co-designed, co-funded, and co-delivered urban agendas have more actual city-shaping potential than individual government actions. This governance alignment has not only sparked a revitalized urban vision in Darwin, but it has also established the institutional framework for long-term, inclusive, and climate-responsive change.

25.2 Climate as Design Logic: Urban Form, Shade, and Tropical Resilience

Darwin's revitalization within the City Deal framework is a necessity for climatic adaptability, not just an urban makeover. The success of any revitalization in a city where heat, humidity, and the severity of seasonal downpours control daily life depends on architectural answers that are closely matched with environmental conditions. Here, climate is not an afterthought; rather, it is the beginning, influencing how buildings touch the ground, how streets are laid out, and how people move, relax, and congregate in public spaces.

The City Deal acknowledges that templates borrowed from temperate places cannot be used to build Darwin's future. Rather, a new lexicon of tropical urbanism is developing, one that incorporates cooling techniques, water management, permeability, and shade into the city's structure. Rethinking streetscapes is at the heart of this. The new interventions concentrate on enlarging walking zones, incorporating significant shade infrastructure, and narrowing carriageways in areas where large roadways predominated and pathways were frequently exposed and isolated. Awnings, canopy trees, and shaded arcades are now essential features that determine how liveable and usable a public area is.

Public buildings and precincts are being redesigned with an emphasis on tropical resilience in addition to these microclimatic enhancements. Architectural elements that are being prioritized in both new and modified buildings include ventilated corridors, elevated floorplates, movable façades, and deep overhangs. In keeping with local culture, the goal is not only to lessen dependency on mechanical cooling systems but also to promote a more laid-back and approachable indoor-outdoor urban lifestyle. Instead of forcing people to adapt to situations that disregard their context, urban design is being shaped to the rhythms of the climate.

Water is viewed as both a limitation and a design opportunity in public infrastructure. Water-sensitive urban design is crucial for drainage as well as for forming the urban space itself because of the heavy rainfall and seasonal storm surges. Permeable pavement, bio-retention systems, engineered wetlands, and swales are being used to manage water in ways that improve ecological and aesthetic appeal. Water features also offer ambient cooling and aesthetic appeal in hotter areas. As part of Darwin's vision, water infrastructure will be incorporated into the city's whole experience, turning it from a technical consideration into a tool for placemaking.

The Civic and State Square Master Plan's execution is perhaps the most notable example of how expansive civic space may serve as a testbed for climate-resilient design. In order to create a readable and welcoming civic precinct, the master design calls for shaded event areas, native vegetation, intelligent lighting, and a high degree of material permeability. Crucially, this area is intended for daily usage and is not only for ceremonial purposes. Heat relief, ventilation, and shelter are thoughtfully layered to allow for comfortable occupancy during periods of high heat.

A rising awareness of thermal comfort is also shown in the use of lightweight and adaptable urban furniture, integrated misting systems, and thoughtful building and seating area orientation. In order to guarantee that people genuinely use the areas being developed, these details—which are sometimes disregarded in expansive urban plans—are essential. Comfort, which in turn depends on the microclimatic calibration of urban form, is the key to liveability in the tropics.

Creating a city that welcomes and celebrates its surroundings rather than just coping with them is the overarching goal of these actions. Residents and visitors are encouraged to stay linked to the outdoors by Darwin's changing urban shape, rather than withdrawing indoors or depending solely on high-energy cooling technologies. It suggests a symbiotic, rather than antagonistic, tropical city where design, function, and environment all work together to promote a resilient, inclusive, and laid-back urban living.

This strategy, which is based on placemaking and pragmatism, establishes a significant precedent. Darwin, the capital of northern Australia, serves as a test site for how climate-smart architecture might transform infrastructure and public areas nationwide. The City Deal is a real-world illustration of how urban shape can be re-evaluated in reaction to the air, light, water, and heat that determine its future, rather than in an abstract manner.

25.3 The New Everyday: Placemaking, Mobility, and Cultural Urbanism

Reviving the city's everyday experience—how people travel through it, congregate there, and connect with it—is at the core of Darwin's transformation under the City Deal. The goal of this regeneration, which goes beyond infrastructure and buildings, is to create spaces where meaningful life emerges in the public sphere. The city is reclaiming its identity as a vibrant (Cheshmehzangi 2021), inclusive, and culturally rich tropical capital by using a placemaking strategy that is both strategic and sensitive.

A key component of this reimagining is the improvement of public spaces. Creating lively, dynamic public spaces that encourage visitors to stay rather than just pass through has been a major focus of the Darwin City Deal. Shade, comfort, and communal use are being given priority in the redesign of Civic Park, State Square, and the waterfront promenades. These areas are now viewed as social and cultural hubs—places of celebration, belonging, and daily rest—rather than just as decorative additions. In a climate where the environment is strongly felt, the focus on permeability, tactile materials, and integrated planting systems enables a sensory interaction with location.

Mobility is a key element of this municipal approach. The city center's conversion into a more accessible, bikeable, and pedestrian-friendly area reflects a broader goal of balancing urban activity and reducing reliance on cars. To facilitate multimodal access, important corridors are being renovated to include better transit stops, protected bikeways, and shaded pedestrian walkways. These improvements make daily travels more pleasant, readable, and integrated into the city's life. They are not just about mobility, but also about experience.

Darwin's perspective on mobility also demonstrates an awareness of inclusivity. The requirements of families, culturally varied communities, the elderly, and those with disabilities are taken into account in the city's spatial renovations. By doing this, they pave the way for a more democratic urban experience, where space is used for public purposes as well as being public in name. A public environment that prioritizes accessibility, care, and social interaction is shown by broad walkways, easy wayfinding, rest areas, and seamless transitions between areas.

A key component of this change is cultural programs. The city's changing identity is becoming more and more reliant on events, festivals, and activations, especially those that highlight Indigenous culture and regional innovation (Cheshmehzangi 2016). Efforts to incorporate cultural tales into performative and physical form have coincided with the revamping of civic places. Darwin's urban future is being firmly anchored in the rich heritage of Country through the use of public art installations, storytelling components, and partnerships with Larrakia custodians. This city aims to highlight its existing culture rather than attempting to impose it on top of it.

Indigenous involvement plays a particularly important function. Partnership with the Larrakia people has been emphasized in the City Deal process, not only as stakeholders but also as knowledge holders and custodians. This change is crucial. It transforms placemaking into an act of recognition, continuity, and shared ownership that goes beyond simple design. places are becoming both social and cultural learning places because to initiatives like dual naming, Indigenous-led design contributions, and interpretative tactics.

The city is also attempting to cultivate smaller-scale urban events that influence the "new everyday". Shaded gathering places, parklets, local markets, and adaptable public spaces all encourage the kind of unplanned and unstructured activity that gives urban life character. In an area where loneliness and transience have historically been problems, the establishment of these "third places"—neither home nor workplace—contributes to social resilience.

In conclusion, Darwin's regeneration aims to disclose and enhance the identity it already possesses rather than imposing a new one. The city is extending an invitation to its citizens to live there more thoroughly and purposefully through cultural urbanism, targeted placemaking, and inclusive mobility. In Darwin, the ordinary is evolving into something greater, influenced by continuity, community, culture, and climate. This type of urban capital is unique in that it is designed to foster community, sharing, and well-being in addition to growth.

25.4 Lessons from the Edge: Regenerating Cities Through Purposeful Collaboration

The Darwin City Deal is a powerful example of how multi-level, purpose-driven cooperation can spur revitalization in smaller capital cities. Geographically remote, climatically distinct, and culturally complex, Darwin, the capital of the Northern Territory and Australia's northernmost city, functions at the periphery of many national discussions. Its renewal is both essential and exemplary, nevertheless, because of this peripheral situation. Cities on the edge—socially, economically, and climatically—can realign themselves with clarity of purpose, coordination of governance, and respect for local uniqueness, according to the lessons learned from the City Deal process.

The Darwin City Deal is fundamentally a lesson on collective government. A single vision for regeneration that transcends jurisdictional borders has been made possible by the tripartite relationship between the City of Darwin, the Northern Territory Government, and the Australian Government. In addition to ensuring funding predictability, this integration of strategic priorities across infrastructure, climate adaption, economic diversification, and liveability has also brought clarity to decision-making. This model illustrates how coordination among levels of government can enhance the potential of scarce resources and decentralized planning for smaller capital cities that are frequently torn between state and federal goals.

The incorporation of climate logic into the urban agenda is another important lesson. Darwin's revitalization is about rethinking what a city needs to be in a tropical setting, not only about enhancing facilities and infrastructure. By putting climate responsiveness at the center of planning and design, from water-sensitive landscapes and shaded corridors to energy-efficient building typologies, the City Deal shows how urban resilience must rise to new heights for regeneration in hot climates. Crucially, this climate-urban integration is viewed as a major design factor influencing daily life rather than as an afterthought or a technical solution.

Co-benefits to the community are also essential to Darwin's methodology. The Darwin City Deal prioritizes both civic and economic benefits, despite the fact that large infrastructure projects are sometimes criticized for being top-down or extractive. A city that aims to support not only tourists or investors but also the daily lives of its citizens is reflected in the improvement of public areas, the inclusion of cultural programming, and the dedication to Indigenous participation. Regeneration is certain to foster public trust and provide evident, shared value among various population groups thanks to its emphasis on human-centered outcomes.

Furthermore, the Darwin City Deal serves as an example of how city accords can transition from lofty aspirations to practical implementation. There have been difficulties along the way; Darwin's revival has need for robust political will, flexible planning techniques, and unambiguous accountability. However, what distinguishes this scenario is its understanding of site specificity, namely that Darwin's transformation cannot replicate models from high-density, temperate capitals. The city is a singularly local, yet nationally noteworthy, story of revitalization because its planned answers are suited to its demographic scale, cultural setting, and climatic extremes.

A roadmap for other edge cities struggling to be viable, dynamic, and climate-ready can be found by considering Darwin's path. The lessons are clear: regeneration necessitates social legitimacy, contextual intelligence, and governance alignment in addition to design quality. Through the Darwin City Deal, a smaller capital city has demonstrated how, not in spite of its peripheral location, but precisely because of it, strategic collaboration, intentional placemaking, and cultural appreciation can unlock new forms of urban vibrancy.

References

Carson D, Schmallegger D, Harwood S (2010) A city for the temporary? Political economy and urban planning in Darwin. Australia. Urban Policy Res 28(3):293–310

Cheshmehzangi A (2016) Temporary and temporality: Public realm regeneration through temporary events. J Urban Regenerat Renewal 10(1):58–72

Cheshmehzangi A (2021) Urban memory in city transitions: the significance of place in mind. Springer, Singapore

City of Darwin (2024) City of Darwin seeks to develop place and liveability plan, Available from: https://www.darwin.nt.gov.au/council/news-media/city-of-darwin-seeks-to-develop-place-and-liveability-plan

City of Darwin (nd) Darwin city deal, Available from: https://www.infrastructure.gov.au/sites/default/files/migrated/cities/city-deals/darwin/files/darwin-city-deal.pdf

Hu R (2019) City deals. In: From Turnbull to Morrison: understanding the trust divide, vol 100, p 7

Kavaarpuo G, Oppong-Yeboah NY, Vuin A (2022) Darwin: towards the sustainability of the Larrikin of Australian capital cities. Cities 120:103457

Novalia W, Suwarso R, Nurdin I (2024) Connecting place and multilevel governance for urban river restoration. Territory, Polit, Govern 1–26

Oakley S (2011) Re-imagining city waterfronts: a comparative analysis of governing renewal in Adelaide, Darwin and Melbourne. Urban Policy and Research 29(3):221–238

Chapter 26
Village Within the City: Glebe Hill and the Revival of Everyday Urbanism in Hobart

26.1 Returning to the Neighbourhood: Framing Glebe Hill as a Contemporary Urban Village

Glebe Hill Village is a conscious attempt to rethink suburban life in Hobart, moving away from the sprawling, automobile-dependent development of the past and toward a more compact, human-scaled, and interconnected urban future (Hamnett and Freestone 2000). The construction, which is located in the rapidly expanding suburb of Howrah on the city's eastern edge, is a prime example of a new trend in urban infill thinking. Instead of viewing suburban expansion as a dispersion of houses into low-density sprawl, Glebe Hill exemplifies the characteristics of a modern urban village (Greater Hobart Committee 2022): it is multipurpose, walkable, and built to accommodate daily routines.

At the core of this change is the notion that proximity, accessibility, and the development of common areas—rather than just density—must characterize urban life. Glebe Hill Village was designed as a neighbourhood-scale hub where housing, retail, green infrastructure, and transportation all come together in a well-planned setting (Greater Hobart Committee 2022). Although the scale is small by city standards, its aim is in its coherence—an integrated destination that provides social interaction, local services, everyday necessities, and a revitalized sense of place.

The idea of "20-min neighbourhoods", or areas where locals can get the majority of their daily necessities in a short walk, bike ride, or local transit journey, is at the heart of Australia's planning movement. This model assumes special importance in Hobart's situation. Because of the city's low-rise profile, sparse public transportation system, and growing growth demands, creative solutions that strike a compromise between compactness and liveability are needed. As a sort of pilot, Glebe Hill fills this void by creating a new suburban hub that enhances the functionality of the eastern suburbs without erasing their unique identity.

The site itself plays a crucial role in this narrative. Glebe Hill, which is situated between Rokeby and Howrah and close to important thoroughfares like Rokeby Road

A. Cheshmehzangi, *Australia's Urbanity*, Urban Sustainability,
https://doi.org/10.1007/978-981-95-3881-2_26

and the South Arm Highway, is at a strategic location that is both close enough to the city center of Hobart to maintain a self-sufficient hamlet and far enough away to be a part of the larger metropolitan network (City of Clarence, n.d.). The precinct anticipates future housing and demographic shifts in addition to reacting to the local population expansion. The development avoids the dispersed results of conventional suburban extensions by focusing activity in one concentrated area and substituting it with a robust spatial anchor.

This goal is supported by the site planning and architectural approach. Glebe Hill Village is intended to be a neighbourhood that encourages continuity with its surrounding setting rather than a stand-alone shopping mall or housing development. Its design prioritizes permeability, promotes pedestrian access from nearby streets, and produces a recognizable framework where residential, commercial, and green areas develop in connection to one another. Its core element is its emphasis on "urban stitching", or re-connecting formerly underutilized land with its urban surrounds.

The ideals ingrained in Glebe Hill Village's design are what set it apart, not just the utilitarian combination it provides (Greater Hobart Committee 2022). It promotes localism as a framework for future resilience rather than as nostalgia for the past (Cheshmehzangi 2021). A supermarket, specialty shops, health and wellness services, and community facilities are all part of the well-planned area. These are essential to the urban logic and are not only incidental to the housing. Glebe Hill's people are empowered by their close proximity rather than being reliant on far-off metropolitan centers, as part of a broader system of mutual assistance.

Additionally, the initiative redefines what a suburban amenity is. In order to promote spontaneity and neighbourly interaction, residential areas are integrated with commercial and social uses rather than being isolated from them. Taking a stroll to get groceries turns into a chance for casual interactions. Public areas can be brought to life by local activities. With the help of landscaping, lighting, and scale considerations, the streets serve as platforms for shared experiences rather than just transporting automobiles.

Planning-wise, Glebe Hill Village is a valuable example of how smaller towns might use urban village models without having to rely on pre-made templates (Greater Hobart Committee 2022). It is a uniquely Tasmanian answer to the urbanity challenge rather than a replica of inner-city Melbourne or Sydney. It asserts a distinct break from the car-dominated suburban norms while reflecting the scale, climate, and cultural traditions of Hobart.

Glebe Hill Village essentially offers a fresh perspective on what neighbourhoods can be in the twenty-first century. It serves as a reminder that major precincts and city centers are not the only places with vibrant urban areas. It can come from the periphery—from suburban locations that have been thoughtfully redesigned with an eye toward public benefit. Therefore, this initiative is not just a significant step in Hobart's eastward expansion, but it also serves as a model for how Australian suburbs might recover their sense of civic purpose.

26.1.1 Local Logic, Urban Reach: Setting the Ground for a New Village Model

Glebe Hill Village was not created overnight. Its emergence was influenced by several factors, including changing expectations regarding suburban expansion in Hobart, strategic urban policy, and demographic momentum. The demand for additional self-contained and liveable suburban centers has grown more pressing as eastern Hobart continues to see continuous population expansion due to affordability concerns, lifestyle migration, and shifting household choices. By grounding its planning in local logic and expanding its reach as a repeatable example for neighbourhood-scale urbanism (Parker and Street 2015), Glebe Hill responds to this request.

The location of Glebe Hill Village is in the established residential suburb of Howrah, which is situated at the nexus of regional service need and urban infill opportunity. The project was intended to serve as a new civic and residential center for the eastern shore as a whole, in addition to being a commercial center. By doing this, it embodies the ideas of comprehensive neighbourhood planning, providing a variety of uses, walkability, social areas, and living close to one another in a cohesive setting.

The project's development required the scaffolding supplied by state and local planning policies. Over the past ten years, Tasmania's planning reforms have placed a greater emphasis on promoting infill rather than expansion, especially in places with excellent accessibility and existing infrastructure. Glebe Hill's planning approvals were made possible by the Clarence City Council's support for walkable, mixed-use zones inside its growth corridors. This development stands out due to its village-scale placement, which is both dense and varied enough to sustain meaningful urban activity while still being tiny enough to stay a part of its suburban environment.

The precinct's viability was based on the idea of retail integration. With the help of smaller specialty stores and service providers, a full-line supermarket generates the foot traffic and economic gravity required to support other operations. However, Glebe Hill's commercial core was created with a civic edge in mind, in contrast to conventional suburban shopping centers. Streetscapes are lively, entrances face outward, and the surrounding public space is designed to promote walking, loitering, and gathering. One of the most important steps in reinventing the suburban experience is the transition from inward-facing commerce to outward-facing urbanism.

In terms of housing, Glebe Hill brings in higher-density housing types in an area that has traditionally been dominated by single-family homes on expansive lots. The development design includes townhouses and medium-rise flats that take advantage of being close to parks, transit, and services. This lessens reliance on cars and increases land-use efficiency while creating a living space that appeals to a wide range of inhabitants, from young families and professionals to elderly downsizers. Crucially, this density is integrated into the precinct's spatial logic and is neither concealed nor marginalized, demonstrating the conviction that suburban intensity and amenity can coexist.

Additionally, neighbourhood-scale infrastructure is essential to the overall coherence of the precinct. In addition to serving locals, thoughtfully designed bike routes, pedestrian walkways, green spaces, and modest meeting areas have been incorporated into the existing urban fabric. As a node in a larger network, the precinct is intended to attract residents from nearby suburbs and strengthen ties within the community at the hyperlocal level.

Alignment with current sustainability goals affects Glebe Hill Village as well. It echoes larger principles in climate-responsive planning with its small footprint, variety of uses, and focus on environmental integration. The precinct helps to lower emissions and encourage healthier urban living by putting services closer to people's homes and improving non-car mobility options. These results are not accidental; rather, they are the result of a purposeful planning framework that places a high value on cross-sector coordination, long-term viability, and place quality.

In conclusion, more than only market need or development opportunity led to the establishment of Glebe Hill Village. It is based on a strategic re-evaluation of what suburban development can entail, community need, and policy foresight. Projects like these provide a concrete road forward as cities like Hobart attempt to control expansion without sacrificing identity or liveability (Cheshmehzangi 2015)—one that functions at a size that people are familiar with but has an ambition that extends well beyond the local.

26.2 Architecture, Access, and Everyday Infrastructure

The ideals of a compact, linked, and community-oriented urban environment come together in a purpose-designed zone at Glebe Hill Village, marking a turning point in Tasmania's suburban development history. Situated at Howrah, roughly 10 km east of Hobart's downtown, the project was thoughtfully designed to address the expanding population demands of the eastern coast and support the transition to more environmentally friendly suburban forms. Glebe Hill Village, the first new shopping center constructed in Greater Hobart in more than ten years, was intended to be a neighbourhood hub with integrated housing, a first-rate public space, and integrated transit links. Its approaches to architecture and urban planning demonstrate a careful consideration of how people move, engage, and live in their daily surroundings.

A full-line Coles supermarket serves as the project's focal point, and over 20 specialized tenancies, including cafes, retail stores, and health services, round out the 5.5-hectare plot of land. The architectural approach to Glebe Hill Village deviates from the conventional inward-facing retail mall typology by turning outward, orienting shopfronts and facades toward a centrally located landscaped plaza and the adjacent streetscapes (Greater Hobart Committee 2022). In order to promote transparency, comfort, and human-scale connection, the architectural language uses warm natural materials, timber panelling, and transparent facades in a modern yet understated style. The low-rise layout of the built form ensures physical permeability and visibility while preserving a friendly appearance for the nearby residential neighbourhoods.

The project's focus on pedestrian mobility and intelligibility is among its most notable aspects. Shaded lounging places and landscaping soften the transitions between built and open space, while a system of pathways links the central plaza with nearby roadways and parking lots. Retail frontages activate the major pedestrian spine, which leads toward public spaces and gathering places and encourages social interactions and informal use in addition to transactional visits. In order to ensure the site's long-term sustainability, the architects, in collaboration with urban designers and landscape consultants, have given careful consideration to micro-mobility infrastructure, such as safe bike parking, well-marked pedestrian crossings, and future electric vehicle charging facilities.

The precinct is car-accessible, but it also encourages walkability and multimodal integration to lessen reliance on cars. Because of the development's better access points from Rokeby Road and surrounding residential streets, locals may easily get there on foot or by bicycle. Adjacent to the site are bus stops that connect Glebe Hill Village to Shoreline, Rosny Park, and the central business district of Hobart. The precinct's layout and advantageous position enable it to serve as a transit-adjacent urban hub, which is an uncommon but increasingly desired goal for suburban Hobart. The precinct design's smooth incorporation of transport corridors facilitates the development of 20-min neighbourhoods throughout the region and is consistent with Tasmania's larger movement towards smaller urban footprints.

The site's everyday infrastructure and architectural performance are integrated with sustainability. In addition to waste management zones, rainwater harvesting systems, and energy-efficient lighting, the store anchor's outside has been landscaped with drought-resistant and native plants. The site's ecological sensitivity and design intelligence are reinforced by permeable paving, bioswales, and shaded pedestrian walkways, which help manage rainwater and provide climate comfort. Despite their modest size, these components work together to create a strong basis for ecologically conscious suburban life.

Importantly, the public areas and physical form have been planned to support future expansion. Zoning flexibility for medium-density residential development, including townhouses and low-rise apartments, in the adjacent properties, while the initial phases concentrate on retail and services. Because of its flexibility, the village concept can expand to meet the needs of the community without needing to be redeveloped in the future, promoting long-term urban resilience. Additionally, it implies that the harmony between everyday service access, employment, and lifestyle can be continuously adjusted throughout time, maintaining its applicability despite changes in the population.

The precinct's spatial layout was thoughtfully planned to facilitate fair access and easy navigation. Passive surveillance, universal access, and movement flows are taken into account while placing important entrances, walkways, and civic landmarks. Families, senior citizens, and individuals with disabilities benefit greatly from the reinforcement of user orientation and safety provided by distinct sightlines, well-defined thresholds, and a variety of material palettes.

The layout and purpose of Glebe Hill Village serve as an example of how ordinary architecture can be rethought without depending on massive scale or high-rise density. Its effectiveness stems from its focus on form, flow, and function—elements that both ground the precinct in its suburban setting and push the limits of what neighbourhood centers can develop into. It establishes a standard for future Australian suburban developments that aim to create stronger, more cohesive communities by striking a balance between architectural coherence, civic purpose, and environmental stewardship.

26.3 Green Spaces and Social Patterns: Designing for Belonging

The idea of an urban village is realized at Glebe Hill Village not only by the mixed-use structures and retail establishments, but also by the ecological and social structures that support the precinct's daily operations. Since the development is situated on Hobart's growing suburban periphery, where established neighbourhoods and new residential developments collide, integrating communal green space and infrastructure is crucial to fostering a feeling of community, place, and belonging. With biodiversity, daily social contact, and ecological value as equal foundations of its urban identity, Glebe Hill Village's architecture exhibits a strong dedication to placemaking through nature.

The precinct's central green plaza, a substantially proportioned, planted forecourt that serves as both a social center and a functional hub, is one of its most important spatial elements. This plaza, which is surrounded by open lawn spaces, public benches, café seating, and shaded areas, is intended to support a variety of activities, from market pop-ups and community events to informal get-togethers and passive relaxation. It encourages the casual interactions that characterize a sense of neighbourhood by allowing users to stay, mingle, and enjoy outdoor comfort. This central open space is not treated as leftover land but as primary social infrastructure, strategically placed to maximise accessibility and visibility from all edges of the site.

Beyond the main plaza, green design is woven throughout the precinct through tree-lined walkways, smaller landscaped alcoves, and planting zones that act as buffers between parking lots and buildings. Throughout, native and climate-resilient flora has been employed to reflect the local Tasmanian landscape and to improve biodiversity. In addition to being aesthetically beautiful, these plants also provide habitat, urban cooling, and sensory enjoyment—all of which are crucial in suburban areas that run the risk of becoming too sterile or hardscaped. Green infrastructure is repositioned as a fundamental urban layer rather than an afterthought as a result of this project.

Crucially, the district has been planned to seem accessible and inviting to the community at large as well as newcomers. While visible permeability throughout the property promotes a sense of safety and inclusion, path networks seamlessly connect with nearby roadways and planned future residential zones. The planning of seating, shade structures, and public amenities has taken into account a wide range of demographics, including families, children, the elderly, and those with impairments. Long-term attachment and community resilience are fostered by the site's inclusive public space, which promotes social variety, impromptu interaction, and frequent use.

The site's shift from the periphery to the center is also sensitively reflected in the design strategies. By incorporating community-focused elements early on, such kid-friendly areas, accessible restrooms, and adaptable event spaces, the precinct establishes itself as a permanent home rather than a transient retail location. Whether it's frequent supermarket excursions that also serve as social gatherings or weekend walks that end with a shared use of green space, these components aid in establishing patterns and rituals into everyday life.

Green spaces in Glebe Hill Village have ecological benefits beyond their social ones. They help manage rainwater, enhance air quality, and improve the built environment's environmental performance. Permeable surfaces and bioswales are two examples of water-sensitive urban design (WSUD) elements that are subtly included into the site's landscape plan, enabling the precinct to work in balance with its natural systems. The perception that the precinct is a part of the area and a broader living system is strengthened by this ecological layering.

Glebe Hill Village develops into a thoughtfully planned urban fabric where nature, community, and architecture converge rather than merely a shopping mall or housing hub. It serves as an example of how even small-scale urban development may actively promote a sense of belonging by acknowledging the value of green space in society and the influence of a well-planned public area (Greater Hobart Committee 2022). The lessons learned from Glebe Hill suggest a more grounded and comprehensive approach to placemaking as suburban areas continue to grow denser and more diverse. This approach should honour the common spaces that enable the unhurried rituals of neighbourhood life.

26.4 Learning from Localism: Small-Scale Urbanism with Big Impact

Glebe Hill Village provides a pertinent example of how purpose-driven, small-scale urban development can have more significant and long-lasting results for expanding cities. This small but significant project on the eastern edge of Hobart shows how careful consideration of community rhythms, ecological integration, and local requirements can create an urban area that seems both deeply rooted and modern. While megaprojects, skyline changes, and large-scale infrastructure are often the focus of discussions about urban growth (ABC News 2025), Glebe Hill serves as a reminder that well-executed minor interventions can produce some of the most long-lasting urban change.

Fundamentally, the precinct embodies a belief in the ordinary—the need of creating urban spaces that facilitate people's daily lives in a dignified and convenient manner. Its green areas, walkable linkages, retail anchors, and housing alternatives are all thoughtfully positioned to promote practical, welcoming, and pleasurable neighbourhood living rather than being ostentatious or avant-garde. In this regard, Glebe Hill Village addresses the pressing demand for urbanism that is both socially and technically regenerative.

The fact that urban regeneration in smaller capital cities does not have to adhere to metropolitan patterns of density and intensity is among the project's most significant lessons. By working with local landscape circumstances, urban grain, and community expectations, it demonstrates that contextual awareness may provide unique, place-based results that are nonetheless replicable and scalable. The village model used here shows that, when combined with careful planning and local accessibility, even low- to mid-rise development can support the more general strategic objectives of infill, housing supply, and less reliance on cars.

The initiative also sheds light on the potential emergence of a new sub-urbanism that shifts away from scattered, car-dominated patterns and toward crowded, connected, and social neighbourhoods. Glebe Hill Village operates as a 20-min neighbourhood in reality (Thornton et al. 2022), not just theory, by integrating housing, services, and the public space into a walking loop. Choice, agency, and social proximity—elements that are frequently absent from more recent suburban expansions—are made possible by its architecture.

Crucially, the project emphasizes the importance of community trust and long-term stewardship. The village was created with the goal of becoming a civic monument and a center of people's life, rather than being thrown in as a stand-alone economic asset. This is particularly crucial in cities like Hobart, where tradition, character, and community identity must be weighed against the fast pace of urban change. Glebe Hill demonstrates that growth does not have to be imposed; rather, it may be an invitation to come together, form connections, and envision a future together.

Glebe Hill Village provides an optimistic and realistic model as Australian cities struggle with the twin issues of housing expansion and urban sustainability. It serves as a reminder that a dedication to people, place, and purpose is more important for meaningful urbanism than grandeur. This little urban community subtly reimagines what a large influence can be through its human scale, local attention, and day-to-day practicality.

References

ABC News (2025) Hobart's urban growth boundary extended to allow fast-growing areas to continue to expand, Available from: https://www.abc.net.au/news/2025-05-23/hobart-urban-growth-boundary-extended-for-development/105325246

Cheshmehzangi A (2015) The reinvention of liveability in public places: Interaction mapping analysis of central Nottingham's improved walkability. J Hum Behav Soc Environ 25(5):426–440

Cheshmehzangi A (2021) Remembering through naming and palimpsests: case study example of the city of Shiraz, Iran. In: Urban memory in city transitions: the significance of place in mind. Springer Singapore, Singapore, pp 243–282

City of Clarence (nd) Glebe Hill Park, Available from: https://www.ccc.tas.gov.au/facility/glebe-hill-park/

Greater Hobart Committee (2022) 30 Year Greater Hobart Plan - Let's Talk, Glenorchy, Available from: https://letstalk.gcc.tas.gov.au/77669/widgets/372697/documents/232103

Hamnett S, Freestone R (Eds) (2000) The Australian metropolis: a planning history, vol 25. Taylor & Francis

Parker G, Street E (2015) Planning at the neighbourhood scale: localism, dialogic politics, and the modulation of community action. Eviron Plann C Gov Policy 33(4):794–810

Thornton LE, Schroers RD, Lamb KE, Daniel M, Ball K, Chaix B, Kestens Y, Best K, Oostenbach L, Coffee NT (2022) Operationalising the 20-minute neighbourhood. Int J Behav Nutr Phys Act 19(1):15

Chapter 27
Reimagining Urbanity: Designing Australia's Next Urban Evolution and Transformation

27.1 From Case to Culture: Redefining Urbanity for an Evolving Nation

This opening section sets the tone for a visionary conclusion by arguing that Australia's urban transformation is not just about physical infrastructure but about cultivating a new urban culture. Based on the case studies covered in this book, it demonstrates how cities are now dynamic, adaptable ecosystems of identity, innovation, and inclusiveness rather than static hubs of growth (Cheshmehzangi 2025a). It considers how people-centered thinking has replaced project-led thinking and how urban design has become a crucial language for the advancement of a country. According to the section, "urbanity" in the Australian context has to be redefined as a civic ethos based on the principles of fairness, diversity, resilience, and common purpose.

Cities in Australia are experiencing a silent revolution that goes beyond zoning laws and infrastructure (Hamnett and Maginn 2016). This is a cultural revolution in how urban life is envisioned, negotiated, and experienced, not just a change in skylines or planning frameworks. The case studies in this book demonstrate how larger concerns like "Who are our cities for?" are increasingly influencing Australia's urban situation. Which values are they expressing? What sort of society do they contribute to creating?

Once used to characterize the physical characteristics of city life, such as density, diversity, and liveliness, the term "urbanity" today necessitates a more complex and nuanced definition. In Australia, civic consciousness is increasingly reflected in urbanity. It is a cultural fabric made of design justice, ecological balance, inclusivity, and everyday belonging in addition to steel, glass, and concrete. The architectural aspirations of mixed-use complexes that put community above consumption, the repurposing of legacy landscapes, and the spatial layouts of new urban areas all reflect this rising attitude.

A. Cheshmehzangi, *Australia's Urbanity*, Urban Sustainability,
https://doi.org/10.1007/978-981-95-3881-2_27

A generational shift can be seen in the shift from project-led to people-centered urbanization. This change can be seen in many of the cities and precincts in this collection, from Darwin's climate-sensitive public realm and Glebe Hill's neighbourhood-scale planning to Fishermans Bend's rebuilt industry-urban interface and Lot Fourteen's blend of innovation and legacy. Each represents a conscious reorientation, with a focus on facilitating significant urban experiences rather than enforcing a constructed answer.

A greater awareness that cities need to be planned with consideration for the people who live there and the histories they hold comes through in these various regions. Australia's urbanity is a dynamic civic culture rooted in the diversity of its multiple identities and the obligations of its public life (Moran 2011), and it can no longer be reduced to metropolitan hubs or economic indexes. This necessitates adopting an integrative perspective that views cities as flexible, living systems and eschewing binary thinking that contrasts the local and the global, tradition and modernity, and form and function.

In this regard, the profession of architecture and urban design has evolved beyond a technical service. It is now a mediator of ethics and a cultural actor. In all of the case studies, we witness architects and urban planners influencing not only the physical environment but also the interactions between institutions, citizens, and the environment. Australia uses design as a language to negotiate its shared goals and make peace with its urban past, whether through integrated green infrastructure, adaptive reuse, or urban infill.

In a settler-colonial setting where identity, memory, and land are still disputed, this is particularly crucial. In Australia, urbanity must take into consideration both incomplete histories and future visions. Recognizing Country, interacting with Indigenous knowledge systems, and incorporating reconciliation into design are all essential components of a genuinely reimagined urban culture. Cities can only become hubs for forward movement and truth-telling at that point. Furthermore, a key theme of urban identity is resilience, in its social, environmental, and economic manifestations. The case studies show how cities are adapting and being open rather than isolating themselves in response to economic uncertainties, demographic shifts, and climate shocks. Every example, from Hobart's innovation-led waterfronts to Epping's suburban development, suggests an urbanity paradigm based on layered resilience and co-benefits.

Lastly, Australian cities are developing into civic possibility labs. They serve as a test site for how design might foster common ideals in a society that is divided. The most potent aspect of urban growth, according to this chapter and this book, is not what we construct but rather how we coexist in it. Therefore, urbanity is the goal of city-building rather than its result. Urban policy, design education, and government must all agree on a straightforward but radical premise in order to realize this vision: a city's generosity, not its size, is what defines it. Australia's urban future hinges on our level of concern as much as our planning.

27.2 Lessons of Urbanity from Micro to Macro: A Checklist for Integrated Placemaking and City-Making

The 25 case studies in this book document the urban evolution of Australia, which shows a significant change in the way cities are planned and constructed. Even if there isn't a single blueprint that can direct every urban future, successful initiatives frequently share strong characteristics, such as design principles, planning frameworks, and cultural insights. These trends create a dynamic checklist for integrated urbanization rather than merely a summary. This is not a list of requirements for regulatory approval or technical compliance. Planners, designers, politicians, and communities are urged to raise the standard for what urbanity might mean and accomplish in Australia and beyond by means of this collection of provocations, which are both aspirational and grounded (see Fig. 27.1).

Perhaps the most pressing and misunderstood imperative of our day is to "Scale with Soul". A flourishing city must be characterized by dignity in addition to density. We observe attempts to humanize scale through the thoughtful articulation of constructed form, diversity of dwelling types, active edges, and visual porosity in projects such as Main Drive Kew, Fishermans Bend, and Epping Renewal. When density is framed around empathy—designing not just for numbers but for the rhythm of daily life—vertical living becomes a form of civic richness rather than compromise. The lesson here is that scale should heighten lived experience rather than flatten it.

"Publicness as a Right" permeates every lively area in this book. The freedom to belong in public space is essential to a meaningful urban life, whether it is

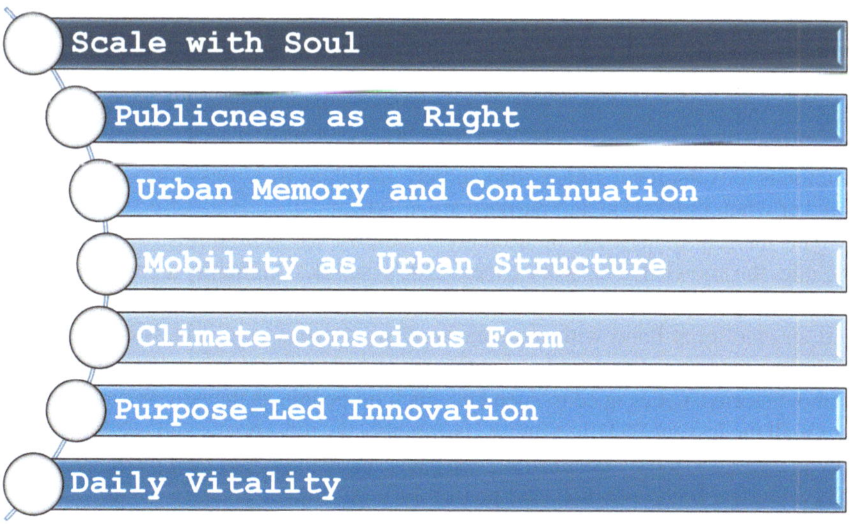

Fig. 27.1 Summary of seven key items in the checklist for integrated placemaking and city-making

seen in Sydney's The Ribbon promenades, Hobart's Macquarie Point gathering grounds, or Darwin's growing civic piazzas. These areas need to be readable, inviting, programmable, and flexible—not just for moving through, but also for remaining. The idea that city life is a shared resource rather than a private one must be reinforced by urban architecture, which must resist enclosure and exclusion.

A more sensitive approach to history is advocated by "Urban Memory and Continuation", which views it as a creative continuation rather than a static preservation. Adaptive reuse offers historic hospitals a new purpose at Lot Fourteen. In Kew, the institutional land and care memories are transformed into a sense of community. Heritage is about establishing a foundation for the future, not about nostalgia. Cities require identity, and identity is constructed in layers. Regeneration needs to be as mindful of the place's history as it is of its physical characteristics.

"Mobility as Urban Structure" reaffirms the significance of transportation as the social and geographical backbone of the city, in addition to its role as infrastructure. Pedestrian networks, bike lanes, and transit corridors are essential to place logic and equity; they are not technical extras. For instance, Glebe Hill Village demonstrates how even suburban settings may encourage walkability and multimodal mobility by combining access with local amenities. Access to opportunities is the essence of good mobility, which needs to be ingrained in every master plan.

"Climate-Conscious Form" is now existential, not elective. We are seeing the early development of cities that view heat, water, and ecological resilience as design factors, from Darwin's shaded boulevards to Hobart and Brisbane's water-sensitive landscapes. The grammar of future-ready urban form now includes green integration, building orientation, permeable surfaces, and shade structures. Given its harsh and diverse environment, Australia's urban development needs to change to become a language of climate responsiveness.

"Purpose-Led Innovation" challenges the idea that economic growth must come at the cost of social or spatial integrity. Innovation precincts like Lot Fourteen and Citiswich show how digital enterprise, sophisticated manufacturing, and research may be integrated into the city's civic life as shared platforms for employment, education, and creativity rather than as corporate silos. It is necessary to curate economic development rather than merely facilitate it. Cities flourish better and quicker when innovation and purpose are in sync.

The unseen framework of urbanity is known as "soft infrastructure", which includes the knowledge networks, social care systems, community areas, and cultural events that give physical space life. This investment is what fosters cohesion, loyalty, and an emotional bond with the area in all prosperous precincts. Missed chances include a neighbourhood without networks, a park without programs, and a street without stories. Cities need to spend as much on people as they do on pavement.

Of all the lessons, "Daily Vitality" is arguably the most poetic and useful. A living city is characterized by bustling streets, pause-inducing edges, and public areas that accommodate both festivities and peaceful times. It takes effort to create, preserve, and curate this energy. Micro-encounters, such as a seating space, a kid's play area, or a café in a shaded corner, are just as important to urbanism as master plans. Even the most technically flawless precinct can feel lifeless without this everyday liveliness, which gives cities their emotional temperature.

These eight principles are a live checklist rather than strict guidelines. They can be used, modified, and contested in a variety of contexts, cultures, and scales. Together, however, they convey a more profound urban ethic: that the future of Australian cities rests not only on their growth but also on their ability to care for people, place, and opportunity.

27.3 Architecture of Belonging: Public Place and the Civic Imagination

This section provides a thorough, introspective look at the philosophical and affective aspects of urban design, examining the ways in which public space and architecture work together to foster a sense of identity, belonging, and civic pride. It makes the case that meaningful cities are more than just intelligent or environmentally friendly. Inspired by multicultural tales, Aboriginal conceptions of Country, and modern place-making, it promotes an "architecture of belonging" that grounds individuals through sensory experience (Cheshmehzangi 2023), cultural resonance, and symbolic continuity. The ethics of space are also covered, including who has the right to construct, live, remember, and imagine.

A true architecture of belonging is the result of the multi-layered processes of democratic involvement, memory-making, and co-creation rather than just artistic skill (Mitchell 2019). In this view, belonging is actively developed rather than passively inherited—through ceremonies that lend built form a collective meaning, intergenerational interactions in libraries, outdoor discussions in plazas, and shared rituals in parks. Ordinary public spaces are transformed into emotionally stirring landmarks by these civic experiences. Architecture thus becomes less about form and function and more about anchoring stories that shape a city's character, both old and new. A stronger, more resilient civic imagination emerges when people can recognize themselves mirrored in the places they live.

In a period of cultural fragmentation, displacement, and climate change, architecture must emerge as a practice of reconciliation that unites people to place, purpose, and one another. The call is to restore, remember, and be anchored, not just to create for functionality or even beauty. Cities can rise above their utilitarian goals and become places of healing and hope by accepting various cultural logics, embracing Country as pedagogy, and honouring varied histories. According to this perspective, architecture unites us in the common endeavour of being human, right here and now, rather than dividing us along class or cultural lines, making the civic sphere a haven of unity.

27.4 Cities as Living Systems: A Vision for Australia and Beyond

Looking ahead, this section makes recommendations for future urban development worldwide, particularly in areas dealing with comparable ecological and socioeconomic issues. It presents the notion that cities are "living systems" that are generative, adaptive, and self-organizing. It describes how cities might be reorganized around life-supporting functions, such as regenerative ecologies, circular economies, and social equity, using Australia's distinctive urban experiments as models. This section urges a rethinking of metrics, moving from GDP and growth rates to joy, access, care, and ecological reciprocity. In the global urban narrative, it presents Australia as a silent pioneer rather than a follower.

We are creating meaning in the cities we construct. Public space is more than emptiness, and architecture is more than shelter; belonging is created in the conversation between the two. The civic imagination starts with how someone feels when they cross a plaza, sit in a courtyard, or stroll down a street rather than with abstract designs. Feeling noticed, accepted, and appreciated is what it means to belong, not just to live there. This section advocates for an "architecture of belonging" that uses sensory perception, cultural resonance, and symbolic continuity to firmly ground individuals.

A common theme among the case studies in this book is that emotional weight must be attached to the physical form. Every successful project, whether it be the community-scale Glebe Hill Village, the adaptive reuse of Kew's institutional buildings, or the shade structures of Darwin's tropical boulevards, does so because it links people to one another as well as to infrastructure. Humans have a need to belong, and cities provide the setting in which that urge is either met or not.

Ethics are as much a part of this type of design as beauty. Deeply political issues surround who is allowed to build, who is allowed to remain, and who is accepted within the city's common area. Inclusion must be ingrained in the procedure and result; it cannot be merely symbolic. We can learn a lot from Aboriginal thoughts on country, where the land is viewed as a living, connected thing rather than just a piece of property on which to build. Space is not neutral in these traditions; rather, it is inhabited by memory, sculpted by care, maintained by community obligation, and told. A truly listening urban design engages in dialogue with the earth rather than merely superimposing form on it.

Extending what we consider to be valid urban narratives is another aspect of this civic imagination. Australia is a multicultural country, and the city ought to reflect that diversity. Diverse forms of identification can be found in Sydney's alleyways, Hobart's municipal center, and Melbourne's resurgent industrial areas. However, this diversity needs to be participatory; it cannot simply aesthetic tokenism. The conditions for people to perceive their stories, languages, and rituals reflected in the surfaces and rhythms of the city must be established by urban planning procedures.

Materiality plays a role here. Light, water, glass, stone, and wood are not neutral tools; they arouse feelings and memories. A stronger feeling of place can be created by the emotional resonance evoked by the smell of native vegetation in a plaza, the sound of footsteps on repurposed masonry, and the coziness of interior spaces lined with timber. The architecture of belonging is about purpose rather than luxury. It establishes places of familiarity and invitation where people feel welcome and their presence is not contingent.

This type of urbanism also has a strong temporal dimension. It respects the past while making room for the present and pointing to unimagined possibilities. New precincts are being planned for continuity as well as utility in places like Adelaide and Brisbane. Time is woven into space with the use of public art, informative paths, memorial design gestures, and changing landscape features. This is crucial for fostering enduring attachment, especially in fast-growing cities that run the risk of deleting the past in their quest for novelty.

The ability of urban architecture to accommodate dreams, however, may be its biggest test. The best public spaces allow for introspection, aspiration, and pause in addition to utility and efficiency. They are places to be as well as places to do. Projects like The Ribbon and Macquarie Point demonstrate how properly planned and executed public spaces can become generative, igniting new civic routines, cultural expressions, and relationships. They serve as a reminder that urban life is co-created rather than merely lived.

In conclusion, the architecture of belonging is the fundamental ethic of city-making and is not an optional layer. It challenges us to live well together in addition to designing well. It pushes us to create spaces where each individual can identify with themselves, share their tales, and take pride in the common urban fabric. The urban future of Australia today need this kind of civic creativity.

27.5 Australia's Urban Century: A Manifesto for the Possible

Australia is entering its urban era, if not its urban century. The question of whether urbanization will shape our future has given way to the question of how and for whom, as cities continue to expand in size, complexity, and impact. We transition from the descriptive to the declarative in this last reflection. This is a manifesto for what might yet be, not a list of what has already happened. It advocates for an Australian urbanism based on morality, creativity, and a common civic goal rather than just growth or practicality.

This book's case studies show a nation undergoing change. From Hobart's post-industrial waterfronts to Darwin's tropical rebirth, from Adelaide's innovation campuses to Sydney's middle-ring suburbs' vertical villages, the designs depict towns discovering their identity. They demonstrate the rise of a brand-new urban culture that is both globally relevant and experimental, place-aware, and intensely local. However, these advancements are not uniform. All too frequently, they depend on isolated partnerships, bursts of financing, or pockets of leadership. A national policy that improves the standard of urban ambition throughout the continent is now required.

Values are where this manifesto starts. Care, innovation, and climate responsibility must be the cornerstones of the cities of the twenty-first century. Our planning must be guided by care, paying close attention to community life, health, equity, and access. Our designs must be shaped by creativity, not as decoration but as a tool for envisioning ambitious possibilities. Every decision must be based on climate responsibility, acknowledging that urban form either contributes to the problem or helps to solve it.

Imperatives flow from these values. Instead of being a specialized duty of local councils, urban policy needs to be rethought as the cornerstone of national governance. Cross-sector cooperation should be anchored by a federal Department of Cities, supported by steady budget, strategic vision, and inclusive frameworks (Aliyeva et al. 2024). Rather than being afterthoughts, planning systems should incorporate design excellence, social value, and cultural responsiveness into their very foundation. Investments in infrastructure must be diverted toward people-first initiatives, such as walkable neighbourhoods, public transportation, educational institutions, libraries, green spaces, and cultural centers that support daily life.

We need to set higher goals at the design scale. Although Australian urban planning and architecture have developed into self-assured, context-driven disciplines, the standard needs to be raised once more. We must insist on structures that inspire, endure, and do more than just serve a purpose. In addition to serving as a buffer between buildings, public areas should be the center of civic life. Instead of being afraid, density could be refined, layered with compassion and constructed with dignity. Place language needs to become more poetic, plural, and participative.

This urban century needs to be a social century, too. Built form and social form are inextricably linked. Relationships must be considered in addition to residents when designing. Soft infrastructure, such as care networks, programming, and unofficial gathering spots, needs to be regarded as just as important as rail and roads. Indigenous knowledge should be prioritized as a source of profound spatial wisdom rather than being consulted on the periphery. The civic imagination must be shaped by displaced communities, aging populations, young voices, and migrant stories. This manifesto also embraces urban possibility or possibilities. It maintains that the cities of the future don't have to resemble scaled-down replicas of the ones that exist today. They have the potential to be more humanistic, expressive, lively, and just. This entails making room for experimentation in the form of innovative housing types, flexible zoning, transitory usage, and cultural hybrids. It entails granting freedom for uniqueness, attractiveness, and local creativity. It entails creating soulful cities as well as clever ones.

And finally, we return to the "pulse of urban life", as the book's sub-title entails. The soul of a city is defined by that ineffable rhythm—of movement and contact, of quiet and intensity, of memory and expectancy. It is what gives urbanity its meaning and ties people to their surroundings. Skylines and our civilization as a whole will be shaped by the cities we create today, including how we design them, who we listen to, and what we prioritize. Building cities means building futures.

So let this be our call: to make the possible irresistible. To shape cities as hubs for shared imagination, resiliency, and belonging rather than just as places to live. The urban century in Australia is not yet over; we have the power to influence it. Let's give it a good shape.

27.5.1 Epilogue: Toward a More Thoughtful Urban Future

One fact emerges from the layers of form, policy, design, and infrastructure as we come to the end of our exploration of *Australian urbanity*: our cities are more than just the locations where we live; they are representations of our values and identity (Jacobs 2012). They are evidence of our collective agency and reflect our aspirations. From Darwin's climatic regeneration to Sydney's vertical reimaginings, a deeper story has been subtly developing in each district covered in this book. This story reframes the city as a moral, cultural, and civic concept rather than just a physical structure. At its core, urbanity is a process, a constant moulding of life via space, time, and relationships, rather than a finished good. Architects and planners are not the only owners (Charnock 2018). Communities, governments, designers, artists, builders, and people who just walk down the street every day are all co-authors of it. Therefore, the future of Australian urbanism depends not just on creative designs or cutting-edge technology but also on our ability to use social agency; that is, our readiness to participate, listen, co-create, and care.

We now have a straightforward yet important question to ask: What kind of people are we invited to become by our cities? This is neither a technological subject, nor is it constrained by infrastructure expenditures or zoning regulations. It comes down to morality and creativity. Are we encouraging inclusion or exclusion in our cities? Isolation or connection? Erasure or memory? Is it dignity or exclusion? The answer lies in the everyday decisions that shape our public spaces, the values embedded in our design frameworks, and the courage with which we navigate difference, history, and change.

Across the 25 case studies presented in this book, three powerful threads have persisted. First, location has the power to influence identity through lived experience as well as symbolic means. The second is how people, cultures, ecologies, and futures can be mediated by careful design. Third, there must be a balance between change and permanence; the best urban futures come from incorporating the past into something fresh, current, and significant rather than from destroying it.

If this concluding meditation contains any call to action, it is that we must dedicate ourselves to the long journey of urban stewardship. This entails planning for belonging as well as advancement (Cheshmehzangi 2025b). It entails putting more emphasis on caring than control, empathic leadership over ego, and regeneration over replication. Real urban progress can be found in the modest success of cities that seem respectable, fair, and alive for everyone, not in world rankings or famous skylines. Australia stands at a threshold moment. Old models are being challenged, and new paradigms are being sparked by the convergence of climate, demography, cultural plurality, and economic change (O'neil and Watts 2015). The obligation is as great as the opportunity. It's time to be bold with your values as well as your design. Now is the moment to embrace urbanity as a civic ethos—a mindset and behaviour that prioritizes people, place, and purpose.

Let this not be the end of the conversation, but the beginning of a new civic imagination. We are already being shaped by the cities we create today. They have the ability to improve, or worsen, our lives as a society. They may be engines of equity or instruments of exploitation. They can either sing or remain silent.

The task before us is clear: to build cities not only for the world we see, but for the society we aspire to be. Let us choose wisely. Let us make a sensible decision. Let us construct urbanity with heart, humility, and purpose. Let's create cities that are profoundly human in addition to habitable and sustainable.

References

Aliyeva A, Balciunaite U, Kamau S, Völler C (2024) Effective cross-sectoral collaboration for strategic sustainable development. Available from: https://www.diva-portal.org/smash/record.jsf?pid=diva2:1873784

Charnock G (2018) Space, form, and urbanity. In: The Sage handbook of Frankfurt School Critical theory, vol. 3

Cheshmehzangi A (2023) City life experiences: lessons learnt for future urban regeneration strategies. In: Mapping urban regeneration: city life experiences in Yunnan, China. Springer Nature Singapore, Singapore, pp. 239–246

Cheshmehzangi A (2025a) Urbanity in flux: the evolving essence of Chongqing. City walks in Chongqing. Springer, Singapore, pp 115–130

Cheshmehzangi A (2025b) Urban planning for the contemporary age: navigating complexities and shaping urban futures. Encyclopedia 5(1):19

Hamnett S, Maginn PJ (2016) Australian cities in the 21st century: suburbs and beyond. Built Environ (1978-), 42(1):5–22

Jacobs JM (2012) Commentary: property and propriety:(re) making the space of indigeneity in Australian cities. Postcolon Stud 15(2):143–147

Mitchell M (2019) Making media with refugee youth in the UK and Lebanon: a practice-based enquiry into co-creation (Doctoral dissertation, Royal Holloway, University of London)

Moran A (2011) Multiculturalism as nation-building in Australia: Inclusive national identity and the embrace of diversity. Ethn Racial Stud 34(12):2153–2172

O'neil C, Watts T (2015) Two futures: Australia at a critical moment. Text Publishing, Melbourne

MIX
Papier aus verantwortungsvollen Quellen
Paper from responsible sources
FSC® C105338

If you have any concerns about our products,
you can contact us on
ProductSafety@springernature.com

In case Publisher is established outside the EU,
the EU authorized representative is:
Springer Nature Customer Service Center GmbH
Europaplatz 3, 69115 Heidelberg, Germany

Printed by Libri Plureos GmbH
in Hamburg, Germany